Impact of Recent Advances in Processing of Ceramic Superconductors

Related titles published by the American Ceramic Society:

Functionally Graded Materials: Manufacture, Properties, and Applications (Ceramic Transactions Volume 76)
Edited by Asish Ghosh, Yoshinari Miyamoto, Ivar Reimanis, and John J. Lannutti
© 1997, ISBN 1-57498-038-6

Case Studies in Ceramic Product Development, Manufacturing, and Commercialization (Ceramic Transactions Volume 75)
Edited by Asish Ghosh, Ron E. Barks, and Basavaraj Hiremath
© 1997, ISBN 1-57498-028-9

Crystal Growth of Novel Electronic Materials (Ceramic Transactions Volume 60)
Edited by R.K. Pandey and Ruyan Guo
© 1995, ISBN 1-57498-003-3

Magnetic Ceramics (Ceramic Transactions Volume 47)
Edited by Bhaskar B. Ghate and John J. Simmins
© 1995, ISBN 0-944904-81-5

Ceramic Transactions
Volume 84

Impact of Recent Advances in Processing of Ceramic Superconductors

Edited by

W. Wong-Ng
National Institute of Standards and Technology

U. Balachandran
Argonne National Laboratory

A.S. Bhalla
The Pennsylvania State University

Published by

The American Ceramic Society
735 Ceramic Place
Westerville, Ohio 43081

Proceedings of the Symposium on Impact of Recent Advances in Processing of Ceramic Superconductors held at the 99th Annual Meeting of The American Ceramic Society in Cincinnati, Ohio, May 5–7, 1997.

For information on ordering titles published by the American Ceramic Society, or to request a publications catalog, please call 614-794-5890.

Printed in the United States of America.

1 2 3 4–01 00 99 98

ISSN 1042-1122
ISBN 1-57498-031-9

Contents

Preface vii

Processing, Phase Equilibria & Single Crystal Growth

High Pressure Synthesis and Properties of Hg, Au, and Carbonate based Superconductors 3
C. Chaillout, S. Le Floch, S.M. Loureiro, J.J. Capponi, J. Chenavas, P. Pordel, E. Gautier, F. Duc, and S. Pachot

Feasibility of Manufacturing Large Domain YBCO Levitators by Using Melt Processing Techniques 13
S. Sengupta, J. Corpus, M. Aganwal, and J.R. Gaines, Jr.

Effect of Y_2BaCuO_5 Particle Size on Undercooling and Solidification Rate of $YBa_2Cu_3O_{7-x}$ 23
F. Dogan and J.D. Reding

A Growth Model for Seeded Melt Processing of Single-Domain $YBa_2Cu_3O_x$ 31
D. Qu, B. Tent, and D. Shi

Primary Crystallization Volume of BSCCO 2212: Applications to Crystal Growth and Melt Processing 41
L.P. Cook and W. Wong-Ng

Pb-Distribution in a Five-Phase (Bi,Pb)-Sr-Ca-Cu-O Assemblage 55
L.P. Cook and W. Wong-Ng

Preliminary Melting Data on the Pb-2223((Bi:Pb)-Sr-Ca-Cu)Phase Under 7.5% O_2. 71
W. Wong-Ng, L.P. Cook, and W. Greenwood

Crystal Growth and Synthesis

XRD Analysis of YBCO Superconductor Films 83
E.J. Peterson, P.N. Arendt, J.Y. Coulter, P.C. Dowden, S.R. Foltyn, J.R. Groves, W.L. Hults, J.L. Smith, and J.O. Willis

Formation of $Bi_2Sr_2CaCu_2O_8$ (Bi-2212) From Metal-Bearing Precursors 93
T.J. Detrie and K.H. Sandhage

Influence of Silver on the $(Bi,Pb)_2Sr_2Ca_2Cu_3O_x$ Incongruent Melting Reaction 105
R.J. Moon, K.R. Trumble, and K.J. Bowman

Influence of Firing Procedures in the Sol-Gel Synthesis of $YBa_2Cu_3O_{6+x}$ Superconductors 117
A. Manthiram, C. Tsang, and C. Ramanujan

Films and Applications

Impact of HTSC-Based Components on Cellular Telecommunications System Performance 129
H. Muller and J.D. Hodge

Normalized Ion Polarizability Additivity Rule and Its Application on HTSC Substrate Materials 139
Ruyan Guo and Amar S. Bhalla

Development of Ceramic Superconductors for Electric Power Applications 157
U. Balachandran

Wires and Composites

Isothermal Melt Processing of Bi2212 Wires and Tapes 175
T.G. Holesinger, J.F. Bingerts, J.O. Willis, and J.Y. Coulter

Coated-Wire-In-Tube (CWIT) Processing of Bi-2223 185
S.E. Dorris, N. Ashcom and N. Vasanthmohan

Stress/Strain Performance of Sheathed Bismuth-2223 Superconductors 195
N. Vasanthmohan, J.P. Singh, G. Khanna, and M.T. Lanagan

Effect of Shock Loading on Critical Current Density of BSCCO/Ag Tapes 207
S. Powell, O. Inal, and U. Balachandran

Interface Modification by Bromine in the Processing of Superconducting Composites 217
S.M. Mukhopadhyay and C. Wei

Superconductor Transitions in Glassy Selenium 231
S.D. Savransky

Index 239

Preface

The discovery of a substantial number of high temperature superconducting oxides and the subsequent intensive research over the past ten years have created major technical breakthroughs and unprecedented scientific excitement. This new frontier of research has captured global attention. Seldom has history witnessed such an extraordinary era that has united scientists and engineers from almost all disciplines. Despite the complexities of the materials, extensive research on the synthesis, processing, characterization, and applications of superconducting oxides has resulted in significant advancements since the discovery of these materials. It is apparent that research and development in these still-growing families of copper oxides will continue for the foreseeable future.

As an indication of the importance of this field, there is a vast and growing body of literature, including presentations, journal articles, and monographs that is devoted to superconductivity research. Each year special symposia and meetings have been organized worldwide to highlight recent activities in the field. The American Ceramic Society (ACerS) recognizes the importance of the high temperature superconductivity research and has continued to support these meetings. The 1997 symposium titled "Impact of Recent Advances in Synthesis and Processing of Ceramic Superconductors" was held May 5–7, 1997, in the Cincinatti Convention Center. The goal of this symposium was to provide a forum to promote interlaboratory cooperation and information exchange among scientists and engineers, and to address the current challenges in the high-T_C superconductivity research. The two-and-half day symposium program included 9 invited talks, 18 contributed talks and 6 poster contributions. Speakers were mainly from the United States, but several were from France and United Kingdom as well.

The symposium featured papers that addressed current processing techniques and methods for high temperature superconductor oxides in various forms, including powder, bulk, thin film, tape and wire. Discussions included novel techniques ranging from the high pressure approach to low temperature sol-gel routes. Because of the remarkable improvement in superconducting properties, the powder-in-tube (PIT) processing method has dominated the tape and wire industry in recent years and has become one of the promising technologies for large scale applications. Presented papers covered a representative range of topics of current interest. These included a review of the development of ceramic superconductors for electric power applications, studies of melt processing of high temperature superconductor tapes, influence of Ag on melt processing, and investigations of the crystallization fields of the high temperature superconductors. The latest developments of thin film technology and dielectric substrates were also highlighted.

The contents of this volume comprise the proceedings of the 1997 superconductor symposium, and the articles cover a large portion of the invited and contributed talks. The papers are divided into 5 parts, corresponding to the following areas: (1) processing and single crystal growth, (2) crystal structure and synthesis, (3) phase equilibria and crystal growth (4) films and applications, and (5) wires and composites. Classification of papers into these categories was not always clear-cut. The order in which the papers appear and the groups into which they are organized may be different from that of their presentation at the meeting. It is hoped that this broad-ranging collection of papers will serve to provide the readers with updated information on the research and developments in the field of high temperature superconductors .

We gratefully acknowledge the assistance of the session chairpersons. We are also in debt to Ms. Sarah Godby, publishing coordinator from ACerS, for her contributions to the editing of this book. Also, special thanks are due to the speakers, authors, manuscript reviewers, and Society officials for their involvment of time and resources.

W. Wong-Ng
(National Institute of Standards & Technology)

U. Balachandran
(Argonne National Laboratory)

A.S. Bhalla
(The Pennsylvania State University)

Processing, Phase Equilibria & Single Crystal Growth

HIGH PRESSURE SYNTHESIS AND PROPERTIES OF Hg, Au, AND CARBONATE BASED SUPERCONDUCTORS

C. Chaillout*, S. Le Floch, S. M. Loureiro, J.J. Capponi, J. Chenavas, P. Bordet, E. Gautier, F. Duc, S. Pachot,
Laboratoire de Cristallographie, CNRS, BP 166, 38042 Grenoble cedex 9, France;

ABSTRACT

The use of high pressure synthesis (2-8 GPa) at high temperature (about 1000°C) has been very successful for the preparation of high T_c superconductors. In some cases, such as the Hg-based cuprates, the obtained phases are not really high pressure ones, but this technique has helped to optimize their synthesis or to stabilize them without any partial cationic substitutions. In other cases, new high pressure phases have been obtained, as the $(Cu,C)Ba_2Ca_{n-1}Cu_nO_x$ family or, more recently, gold-based high T_c cuprates.

INTRODUCTION

In the recent years, the use of high pressure synthesis at high temperature has led to the dicovery of many new high T_c superconducting families (table I). In some cases, like the Hg-cuprates, the new compounds are not high pressure phases, but nevertheless the use of pressure has greatly helped to optimize the synthesis of the samples or to stabilize the phase without partial substitutions for Hg. In other cases, the synthesis under high pressure is the only way to obtain the phases.

In the present paper, after an introduction to the general features of the high pressure synthesis technique used in our laboratory, examples of high pressure phases we have synthesized and studied will be given: the Hg-based phases, the $(Cu,C)Ba_2Ca_{n-1}Cu_nO_y$ ((Cu,C)-12(n-1)n) family and the Au-1212 phase.

Table I. Examples of high T_c superconductors prepared under high pressure

Year	Series	Tc Max	Ref.
1993	$Sr_2Ca_{n-1}Cu_nO_y$	109	1,2
	$HgBa_2Ca_{n-1}Cu_nO_{2n+2+x}$	136	3
1994	$(Cu,C)Ba_2Ca_{n-1}Cu_nO_y$	117	4-7
	$AlSr_2Ca_{n-1}Cu_nO_{2n+3}$	110	8
	$Hg_2Ba_2(Y,Ca)_{n-1}Cu_nO_{2n+4-x}$	80	9,10
	$GaSr_2Ca_{n-1}Cu_nO_{2n+3}$	109	11
	$BSr_2Ca_{n-1}Cu_nO_{2n+3}$	110	12
	$(C,B)(Sr,Ba)Ca_{n-1}Cu_nO_y$	118	13
	$PbSr_2Ca_{n-1}Cu_nO_y$	115	14
1995	$(Cu,S)Sr_2Ca_{n-1}Cu_nO_y$	100	15
1996	$AuBa_2(Y,Ca)_{n-1}Cu_nO_y$	80	16

SYNTHESIS

The synthesis is usually carried out in two steps. The first one consists of the preparation of a precursor from a mixture of basic oxides or nitrates ($Ba(NO_3)_2$, CaO, CuO, ..) heated in air or oxygen at a temperature between 800 and 930°C. The resulting product contains both stoichiometric and non-stoichiometric phases such as $BaCuO_2$, Ca_2CuO_3, Ba_2CuO_{3+x}, $Ba_2Cu_3O_{5+x}$, CuO. The synthesis conditions of the precursor (time, temperature, cooling rate) impose not only the composition but also the proportion between stoichiometric and non-stoichiometric phases and, therefore, determines the average valence of copper. For example, we have shown that, in Hg-based phases [17], this average valence is a determining factor for the synthesis of the different members of the series.

The precursor is then mixed with HgO, or Au_2O_3, or $BaCO_3$, or $BaNO_3$ depending on the prepared phase and placed in a platinum or gold capsule for the high pressure synthesis which is carried out in a belt-type apparatus. The capsule is placed inside a cell made of pyrophillite acting as a pressure transmitter, introduced inside a chamber pressed between two cone-cylinder pistons made of tungsten-carbide. Inside the cell, the capsule is surrounded by an isolating Al_2O_3 cylinder and a furnace made of graphite or molybdenum.

Depending on the desired phase, we apply a pressure in the range of 2-8GPa and at a temperature close to 1000°C. The reaction is usually completed in a short period of time (0.5 to 2hrs) after which the heat source is suddenly turned off and the pressure is released.

Although high pressure synthesis has been very successfull as will be illustrated below, some limitations should be mentioned. First of all, the sample volume is rather small (≤300mg) which may be a drawback in view of powder neutron diffraction experiments. Second, the reproducibility is not always perfect, mainly because of the instabilities of the applied current at longer annealing times. Third, there may be a reaction of the capsule with the sample, which has two effects: for example in the case of Hg-based cuprates, this effect leads to the removal of Hg from the reacting system and the incorporation of gold or the presence of platinum containing phases.

Hg-BASED CUPRATES

$HgBa_2Ca_{n-1}Cu_nO_{2n+2+\delta}$

These compounds are synthesized through the reaction of a precursor prepared as described above with HgO. At ambient pressure, mercury oxide decomposes at 450°C, a temperature at which the precursor is not yet active. This fact imposes the use of confined synthesis methods such as sealed tube techniques. A way to circumvent this difficulty is to apply high pressure in order to increase the decomposition temperature of HgO. This method allows simple solid state reactions to occur instead of heterogeneous processes that happen between mercury vapour and the precursor when the synthesis is carried out in a sealed tube. The first superconducting Hg-based oxide, $HgBa_2CuO_{4+\delta}$, has been synthesized in sealed tubes [18] and this method has been more or less successfull for members up to n=3 but the high pressure synthesis is the only known process until now which permits to get monophasic samples with n≥5 [19].

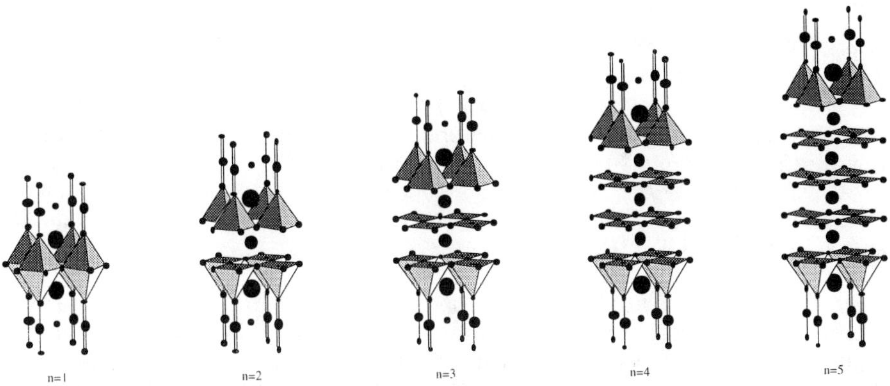

Figure 1: Structure of $HgBa_2Ca_{n-1}Cu_nO_{2n+2+\delta}$ (n=1, 2, 3, 4 and 5)

Figure 1 shows the structural arrangement of the Hg-based cuprates for n=1 to 5. It consists of a stacking of (HgO_δ), (BaO), (CuO_2), (n-1)[Ca, (CuO_2)], (BaO), and (HgO_δ) layers. In these compounds, superconductivity is intimately connected to the occupation of the extra oxygen δ at the center of the Hg mesh. However, for a given T_c, the samples prepared under high pressure and those prepared in sealed silica tubes present a somewhat different δ. This behaviour could be explained by the incorporation of carbonate groups in the place of Hg during high pressure synthesis ; this would provide an appreciable reduction of the Cu valence in the (CuO_2) layers and, therefore, a decrease or even suppression of T_c [20].

$Hg_2Ba_2(Y,Ca)Cu_2O_{8-\delta}$

A double Hg layer sequence, similar to what occurs in the Bi or Tl double layer homologous series ($Bi_2Sr_2Ca_{n-1}Cu_nO_y$ and $Tl_2Ba_2Ca_{n-1}Cu_nO_y$) has been first observed by high resolution electron microscopy [21] as a structural defect in a Hg-1223 crystallite. Later on, $(Hg_{1.5}Pr_{0.3}Cu_{0.2})Ba_2(Y,Pr)Cu_2O_{8-\delta}$ was synthesized by Martin et al. [9] in sealed silica tubes. Almost simultaneously, Radaelli et al. [10] succeeded in synthesizing $Hg_2Ba_2YCu_2O_{8-\delta}$ by using high pressure. This compound became superconducting when Y was partially replaced by Ca with Tc ≈ 80K for about 40% of Ca.

The structure of the n= 2 member of the family as determined from powder neutron diffraction data [10, 23] is shown in figure 2.

By substituting partially Tl for Hg, Tatsuki et al. [22] have been able to obtain, still under high pressure, the n=1 to 6 members of the $Hg_2Ba_2Ca_{n-1}Cu_nO_y$ homologous series, with T_c up to ≈114K.

The formation of superconducting samples from synthesis in sealed tubes has not been reported yet.

One of the interesting properties of $Hg_2Ba_2(Y,Ca)CuO_{8-\delta}$ is a large pressure effect on T_c, with an enhancement of 50K of T_c for a 20GPa applied pressure [24].

OXYCARBONATE CUPRATES

In 1994, we synthesized a new superconducting family in the Ba-Ca-Cu-O system at high temperature and high pressure [4]. Similar results were reported almost simultaneously by two other groups [5,6]. Subsequently, the presence of carbon, forming CO_3 groups, was detected by high-resolution electron microscopy [25] or EELS [7]. The general formula $C_yCu_{1-y}Ba_2Ca_{n-1}Cu_nO_x$ was thus proposed for this family. Four members, with n=3 to 6, have been observed by electron diffraction, but only the n=3 and n=4 members were obtained as bulk phases and characterized from the structural and physical properties point of view.

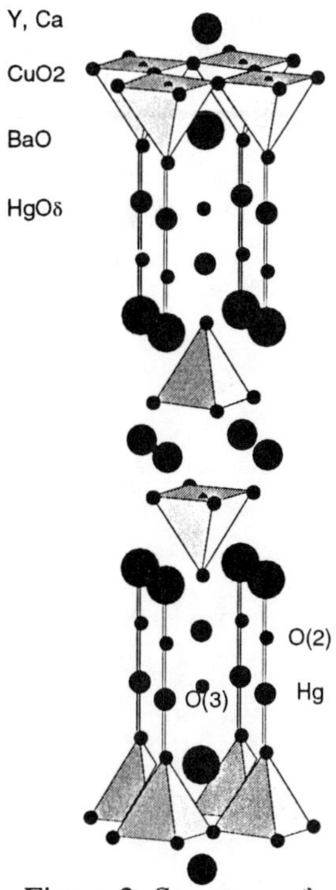

Figure 2: Structure of $Hg_2Ba_2(Y,Ca)Cu_2O_{8-\delta}$ [10]

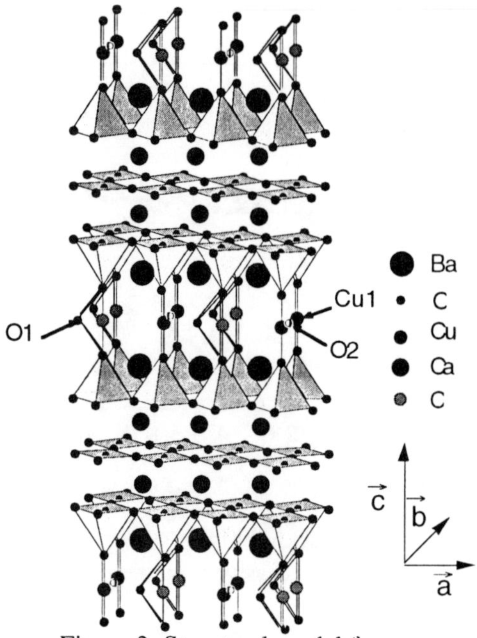

Figure 3: Structural model for $C_{0.5}Cu_{0.5}Ba_2Ca_2Cu_3O_{8+\delta}$

Powder x-ray diffraction patterns were indexed on a tetragonal cell with parameters a=b≈3.86Å and c≈ 14.75 and 18.0 Å for n=3 and n=4, respectively. However, a superstructure corresponding to the doubling of both a and c axes was evidenced by electron diffraction and attributed to the ordering of C and Cu atoms in a 1/1 ratio in the basal plane of the structure. Such a superstructure had already been reported by Hervieu et al. [26] for $Ba_4(Y_{1-x}Ca_x)_2Cu_5CO_3O_{11}$ which can be considered as the n=2 member of the present family.

Due to the presence of unknown impurities and the small volume of synthesized samples, we have not been able to determine the detailed structure of $C_yCu_{1-y}Ba_2Ca_{n-1}Cu_nO_x$ from powder neutron or x-ray synchrotron data. However, it seems that for the n=3 as-prepared samples, oxygen is present in the basal plane of the structure between two Cu atoms along the edge of the cell. This position is similar to the one occupied by the mobile oxygen in $Ba_2YCu_3O_7$. A structural model is given in figure 3.

From a.c. susceptibility measurements, superconducting transition temperatures around 67K and 117K were reported for the n=3 and n=4 as-prepared samples, respectively. A phase with a higher T_c (126K) was also reported to exist in some samples ; but it was not identified, owing to its very small superconducting volume [27]. Recently, Chu et al. reported [28] that this phase might be a Ca rich one, with a formula $Ca_yCu_{1-y}Ba_2Ca_{n-1}Cu_nO_x$.

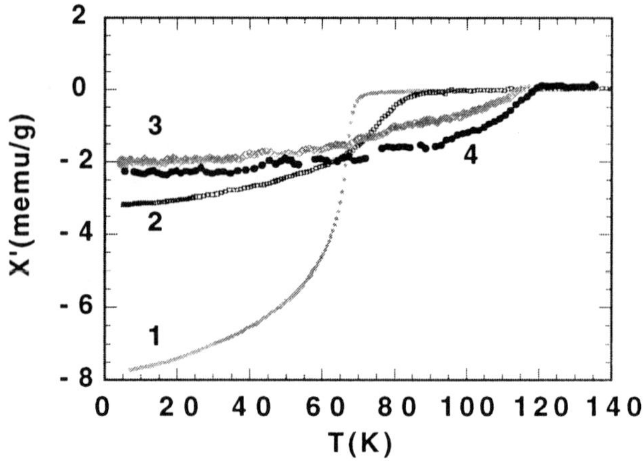

Figure 4: a.c. suceptibilty of a $C_{0.5}Cu_{0.5}Ba_2Ca_2Cu_3O_{8+\delta}$ sample, as prepared (1) and after successive reduction treatments under Ar ((2):2hrs; (3):4hrs; and (4):4.5hrs)

Jaime et al. [27] studied the influence of pressure on the superconducting transition temperatures of the n=4 and n=3 samples. For the former, dT_c/dP was small indicating that the n=4 member is close to optimal doping. On the other hand, for the n=3 member, a strong decrease of T_c with pressure was observed, indicating that this phase may be in an overdoped state. This result motivated us to study the possibility of changing the doping of the n=3 member by either varying the oxygen stoichiometry or replacing divalent Ca by trivalent Y, or substituting N for C. The same kinds of treatments were applied to the n=4 member for comparison.

We have been able to raise the T_c of a $C_yCu_{1-y}Ba_2Ca_2Cu_3O_{8+\delta}$ sample from 67K to 120K by heat-treatments under reducing conditions (figure 4) [30]. This value is higher than that found for the n=4 member, so also in this family the highest T_c corresponds to n=3 as for the other superconducting cuprate families. We have shown that it is possible to vary only slightly the C content around y=0.5, the highest T_c being obtained for the reduced samples with the lowest C content. We tried to substitute partially nitrogen for carbon in our samplesby adding a mixture of $BaCO_3$ and $Ba(NO_3)_2$ to the precursor for the high pressure synthesis. Nitrogen is indeed present in the samples, as evidenced by EELS analyses, and the T_c of the as-prepared samples is somewhat increased, as expected from Cu valence calculations. However, after reduction, the highest T_c is not modified significantly.

Finally, we have been able to substitute up to 10% of Ca by Y ; but this does not seem to be favorable for increasing T_c.

This oxycarbonate series seems very attractive and may prove to be very important in view of applications because, besides exhibiting highest T_c values, it does not contain any toxic element like Hg, Tl, or Bi.

THE Au-BASED PHASE: $AuBa_2(Y,Ca)Cu_2O_7$

When studying the possible substitutions for Hg in Hg-based cuprates, we have demonstrated the possibility of replacing up to ≈ 40% of mercury by gold [30]. The study of the effect of gold incorporation in the Hg-1223 compound showed that T_c and the a and c cell parameters decreased concomitantly with the increase of gold content. This variation was interpreted as due to a gradual overdoping of the CuO_2 layers in the gold-doped compounds, with respect to the undoped ones. Despite many efforts, we were unable to substitute more than ≈ 40% of mercury by gold in these compounds.

Based on the above observations, we have successfully synthesize the new $AuBa_2YCu_2O_7$ compound (Au-1212) by the high pressure-high temperature technique. Assuming the formation of chains of AuO_4 squares in the basal plane of the structure, as for copper in the $YBa_2Cu_3O_7$ compound, a charge equilibrium calculation leads to a 2+ valence state for the copper cations in the CuO_2 planes. We have then gradually increased the doping level by partial substitution of calcium for yttrium up to 40% of Ca. This lead to superconductivity with T_c at ≈82K (figure 5).

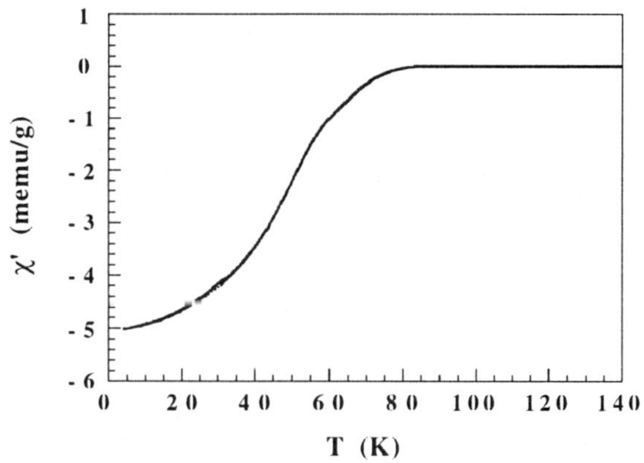

Figure 5: a.c. susceptibility versus temperature for $AuBa_2Y_{0.6}Ca_{0.4}Cu_2O_7$

The structural properties of this system have been investigated by means of electron microscopy and powder x-ray diffraction. The average structure (orthorhombic Pmmm, a=3.8298(5) Å, b=3.8420(5) Å, c=12.111(1) Å for nominal composition 40% of Ca) is very similar to that of $YBa_2Cu_3O_7$, with Au^{3+} cations in square planar coordination sharing corners to form chains running along the b axis (figure 6a).

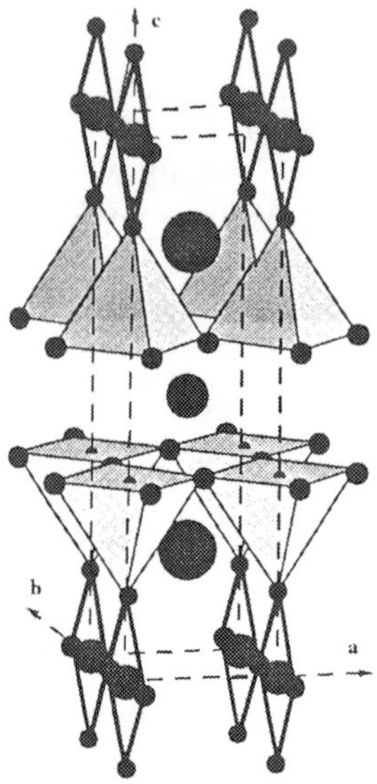

Figure 6a: Structural model for $AuBa_2Y_{0.6}Ca_{0.4}Cu_2O_7$

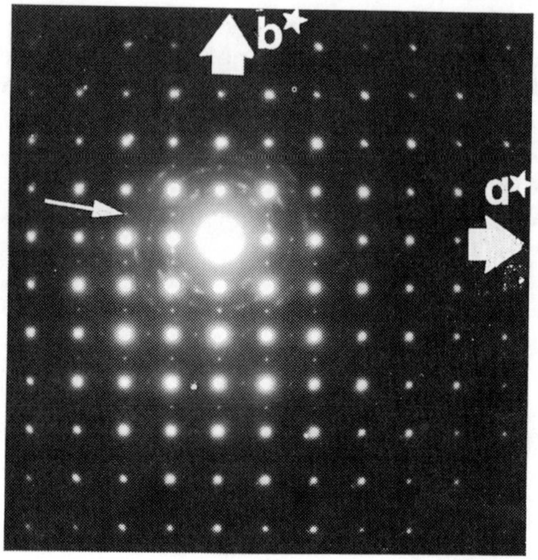

Figure 6b: Electron diffraction pattern taken along the [001] zone axis and showing the superstructure reflections

Electron microscopy reveals the presence of a superstructure with double b parameter (figure 6b), which is attributed to a long range ordered zig-zag conformation of the square chains.

CONCLUSIONS

We have shown, through a few examples, the important role that high pressure synthesis has played in the search for new high Tc superconducting phases or their optimization. In the past few years, most of the discovered homologous series have been indeed synthesized with this technique. Very likely in the future, this will be used to investigate more systems, not necessarily based on CuO_2 planes. One can already mention the spin ladders family $Sr_{n-1}Cu_{n+1}O_{2n}$, that was first obtained during the study of the Sr-Cu-O system under pressure [31].

REFERENCES

[1] S. Adachi, H. Yamauchi, S. Tanaka, N. Nori, *Physica C*, **208** 226 (1993)

[2] Z. Hiroi, M. Takano, M. Azuma, Y. Takeda, *Nature*, **364** 315 (1993)

[3] A. Schilling, M. Cantoni, J.D. Guo, H.R. Ott, *Nature*, **363** 56 (1993)

[4] M. A. Alario-Franco, C. Chaillout, J.J. Capponi, J.L. Tholence, B. Souletie, *Physica C*, **222** 52 (1994)

[5] H. Ihara, K. Tokiwa, H. Ozawa, M. Hirabayashi, A. Negishi, H. Natahuta, Y.S. Song, *Jpn J. Appl. Phys.*, **33** L503 (1994)

[6] C.Q. Jin, S. Adachi, X.J. Wu, H. Yamauchi, S. Tanaka, *Physica C*, **223** 238 (1994)

[7] T. Kawashima, Y. Matsui, E. Takayama-Muromachi, *Physica C*, **224** 69 (1994)

[8] M. Isobe, T. Kawashima, K. Kosuda, Y. Matsui, E. Takayama-Muromachi, *Physica C*, **234** 120 (1994)

[9] C. Martin, M. Hervieu, G. van Tendeloo, F. Goutenoire, C. Michel, A. Maignan, B. Raveau, *Solid State Comm.*, **93** 53 (1994)

[10] P.G. Radaelli, M. Perroux, M. Marezio, S. de Brion, J.L. Tholence, Q. Huang A. Santoro, *Science*, **265** 380 (1994)

[11] E. Takayama-Muromachi, M. Isobe, *Jpn J. Appl. Phys.*, **33** L1399 (1994)

[12] E. Takayama-Muromachi, Y. Matsui, K. Kosuda, *Physica C*, **241** 137 (1995)

13 M. Uehara, M. Uoshima, S. Ishiyama, H. Nakata, J. Akimitsu, Y. Matsui, T. Arima, Y. Tokura, N. Mori, *Physica C*, **229** 310 (1994)

[14] H. Yamauchi, T. Tamura, X.J. Wu, S. Adachi, S. Tanaka, *Jpn J. Appl. Phys*, **34** L349 (1995)

[15] E. Takayama-Muromachi, Y. Matsui, J. Rairez-Castellanos, *Physica C*, **252** 221 (1995)

[16] P. Bordet, S. Le Floch, C. Chaillout, F. Duc, M.F. Gorius, M. Perroux, J.J. Capponi, P. Toulemonde, J.L. Tholence, *Physica C* , **279** (1997)

[17] S.M. Loureiro, C. Stott, L. Philip, M.F. Gorius, M. Perroux, S. Le Floch, J.J. Capponi, D. Xenikos, P. Toulemonde, J.L. Tholence, *Physica C*, **272** 94 (1996)

[18] S.N. Putilin, E.V. Antipov, O. Chmaissem, M. Marezio, *Nature*, **362** 226 (1993)

[19] J.J. Capponi, E.M. Kopnin, S.M. Loureiro, E.V. Antipov, E. Gautier, C. Chaillout, B. Souletie, M. Brunner, J.L. Tholence, M. Marezio, *Physica C*, **256** 1 (1996)

[20] S.M. Loureiro, E.T. Alexandre, E.V. Antipov, J.J. Capponi, S. de Brion, B. Souletie, J.L. Tholence, M. Marezio, *Physica C*, **243** 1 (1994)

[21] G. van Tendeloo, C. Chaillout, J.J. Capponi, M. Marezio, E.V. Antipov, *Physica C*, **223** 219 (1994)

[22] T. Tatsuki, A. Tokiwa-Yamamoto, T. Tamura, X.J. Wu, Y. Moriwaki, S. Adachi, K. Tanabe, *Jpn J. App. Phys.*, **35** L205 (1996)

[23] S.M. Loureiro, C. Chaillout, P. Bordet, J.J. Capponi, P. Toulemonde, P.G. Radaelli, E.V. Antipov, to be published

[24] C. Acha, S.M. Loureiro, C. Chaillout, J.L. Tholence, J.J. Capponi, M. Marezio,M. Nunez-Regueiro, *Solid State Comm.*, **102** 1 (1997)
[25] M. A. Alario-Franco, P. Bordet, J.J. Capponi, C. Chaillout, J. Chenavas, T. Fournier, M. Marezio, B. Souletie, A. Sulpice, J.L. Tholence, C. Colliex, R. Argoud, J.L. Baldonedo, M.F. Gorius, M. Perroux, *Physica C*, **231** 103 (1994)
[26] M. Hervieu, P. Boullay, B. Domenges, A. Maignan, B. Raveau, *J. Solid State Chem.*, **105** 300 (1993)
[27] M. Jaime, M. Nunez-Regueiro, M. Alario-Franco, C. Chaillout, J.J. Capponi, A. Sulpice, J.L. Tholence, S. de Brion, P. Bordet, M. Marezio, J. Chenavas, B. Souletie, *Solid State Comm.*, **97** 131 (1996)
[28] C.W. Chu, Z.L. Du, Y. Cao, Y.Y. Xue, Y.Y. Sun, I. Rusakova, L. Gao, *Philosophical Magazine Letters*, **75** 15 (1997)
[29] C. Chaillout, S. Le Floch, E. Gautier, P. Bordet, C. Acha, Y. Feng, A. Sulpice, J.L. Tholence, M. Marezio, *Physica C*, **266** 215 (1996)
[30] P. Bordet, S. Le Floch, J.J. Capponi, C. Chaillout, M.F. Gorius, M. Marezio, J.L. Tholence, P.G. Radaelli, *Physica C*, **262** 151 (1996)
31 Z. Hiroi, M. Takano, M. Azuma, Y. Takeda, *J. Solid State Chem.*, **95** 230 (1991)

FEASIBILITY OF MANUFACTURING LARGE DOMAIN YBCO LEVITATORS™ BY USING MELT PROCESSING TECHNIQUES

S. Sengupta, J. Corpus, M. Agarwal, and J. R. Gaines, Jr.
Superconductive Components, Inc.
1145 Chesapeake Avenue
Columbus, OH 43212

ABSTRACT

The fabrication of large domain levitators™ for frictionless superconductor-magnet bearings involves synthesis of phase pure powder, single crystal growth, melt processing, and oxygenation. Phase-pure YBCO powders used for fabrication of levitators™ are produced using a low pressure calcination technique. The $Nd_{1+x}Ba_{2-x}Cu_3O_y$ single crystal seed used to initiate the grain growth during melt-processing are produced using a self flux growth method. In order to demonstrate the feasibility of large scale manufacturing of YBCO levitators™, batch sizes of about 90 samples were fabricated in a furnace. A distribution in levitation forces of the levitators™ are observed. Possible factors responsible for this distribution are discussed in this paper.

INTRODUCTION

Melt processing techniques are widely used to fabricate strongly linked $YBa_2Cu_3O_x$. Recently, these materials are being used for developing frictionless superconducting-magnet bearings for energy storage applications[1,2]. Other applications envisioned based on these materials are fault current limiter, current lead, quasi-permanent magnets, etc[1].

In a typical melt processing method, YBCO is heated above its peritectic point where it melts incongruently into Y_2BaCuO_5 and a Ba- and Cu- rich liquid. The semi-solid melt is cooled slowly to obtain aligned grains (or domains) of YBCO. In order to achieve high levitation forces suitable for frictionless bearings, large domain size (that determines the length scale of the induced current loop, d), and high critical current density, J_c, are desirable. One way to increase the domain size is by initiating grain growth by using a seed crystal.

In the presence of a favorable temperature gradient, the seed not only ensures a single nucleation site but also permits controlled orientation of the grains. The seeding technique along with a controlled temperature gradient enables growth of YBCO domains as large as that of the sample size (~ single domain)[4-10].

In this paper we report our results on fabrication of large domain YBCO levitators™ in large batch sizes (~ 90 samples/batch). The YBCO levitators™ are fabricated in near net-shape. The distribution of the levitation forces of the levitators™ are also reported.

FABRICATION OF LEVITATORS™

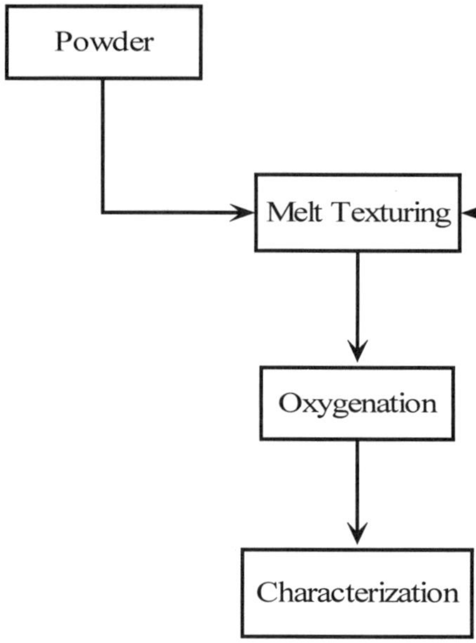

Figure 1: Flow diagram for the levitator fabrication process

Figure 1 shows the flow diagram for the levitator™ fabrication process. The process involves synthesis of phase pure powder, single crystal growth, melt processing, and oxygenation. All of these steps are critical for successful fabrication of levitators™ with desired properties.

Figure 2: SEM photograph of a $Nd_{1+x}Ba_{2-x}Cu_3O_y$ single crystal used for seeding.

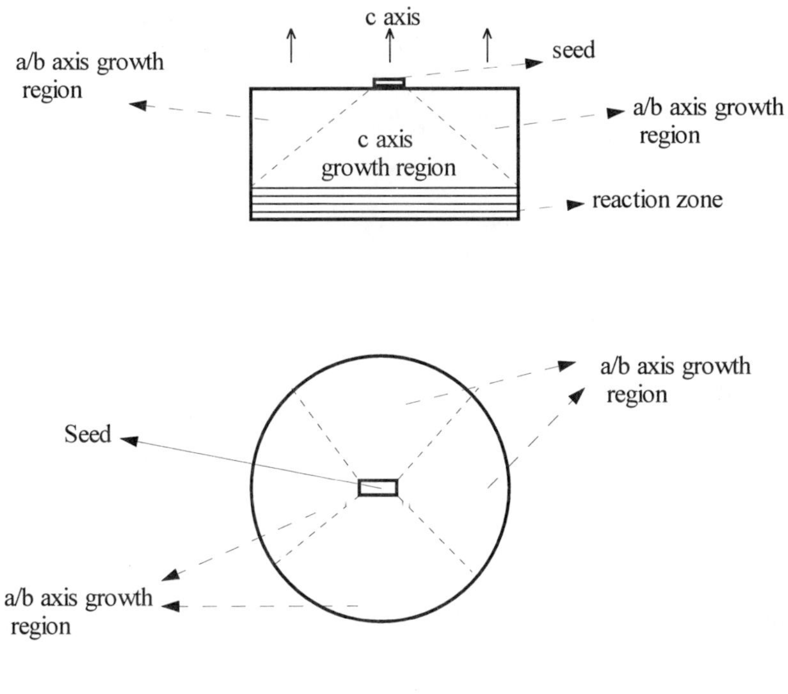

Figure 3: Sketch of various regions in a typical large domain levitator™.

Synthesis of Powder

Phase pure $YBa_2Cu_3O_x$ and Y_2BaCuO_5 are produced by a low pressure calcination technique[11]. The required oxides and carbonates are mixed and ball milled in a proper molar ratio. The resulting powders are then calcined in a low pressure furnace to produce the desired phase. Typically, an oxygen pressure of 2-3 torr is used during calcination. Once the reaction is completed, the vacuum is discontinued and the powders are cooled in an oxygen atmosphere at ambient pressure to fully oxygenate the superconducting powders. Because of the lower processing temperatures, the particle size of the resulting powders is found to be smaller than those obtained by a typical solid state reaction method. Typical particle sizes of the powders obtained by this method range from 1 μm - 5 μm.

Single Crystals Growth

The $Nd_{1+x}Ba_{2-x}Cu_3O_y$ single crystals are produced by using a self-flux growth method[7] (Fig. 2). Specifically, Nd_2O_3, $BaCO_3$ and CuO are mixed in desired ratio by ball milling. The powder is then heated to about 1075-1125°C in a Al_2O_3 crucible and then cooled slowly (~1-5 °C/h) to 1000 °C. The excess flux is then removed and the crystals are cooled to room temperature. The single crystals are then removed and cleaned by an infiltration technique using porous Y_2BaCuO_5 substrate. About 50-150 single crystals are generally produced in a run.

Melt Processing

The YBCO levitators™ are fabricated using a top-seeded melt processing technique. The details have been reported elsewhere[7,9]. Grain growth is initiated by using a $Nd_{1+x}Ba_{2-x}Cu_3O_y$ single crystal as a seed. Nucleation of the $YBa_2Cu_3O_x$ phase at the surface of the $Nd_{1+x}Ba_{2-x}Cu_3O_y$ single crystal occurs after the slow cooling process has started. The crystal begins growing epitaxially in the form of a thin, square platelet, with well developed (100), (010) and (001) planes. The pedestal that develops below the seed is formed by well defined growth steps. These steps are relatively close to the seed but much less pronounced at the edge of pellet. The growth stops when the growth front reaches the reaction layer that forms at the interface between the crucible and the sample (Fig. 3). As reported earlier[9], the reaction layer is usually 1-3 mm thick and consists of a large amount of Y_2BaCuO_5 embedded in Ba-Cu-O phases. Small grains of $YBa_2Cu_3O_x$ and Y-Ba-Cu-Al-Pt-O phases are also observed in these regions.

Figure 4: Photograph of a large domain hexagonal levitator™

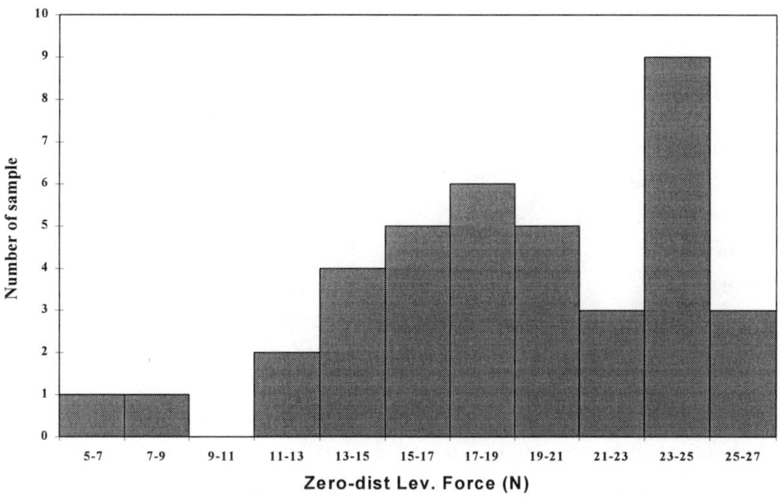

Figure 5: Distribution of Zero-distance levitation force for a batch of 39 levitators™

Oxygenation of Levitators

The levitators™ are oxygenated between 700-450 °C by using a step oxygenation schedule for 120-240 h in a flowing oxygen atmosphere. Generally, about 75-100 levitators™ are oxygenated in a batch. Baffles are used to homogeneously distribute and heat the flowing oxygen gas to ensure adequate oxygenation for every levitator™.

RESULTS AND DISCUSSION

Figure 4 shows a photograph of a typical hexagonal levitator™. These levitators are about 0.45 inch thick with flat to flat separation about 0.7-0.8 inch.

The YBCO levitators™ are characterized by the levitation force on a 12.8 mm diameter and 12.8 mm tall Nd-Fe-B magnet attached to 6.4 mm of Fe. The details of the measurement are reported elsewhere[7,9]. The levitator™ is first zero-field cooled to 77 K by using liquid N_2. The levitation force is measured as the magnet is moved towards the levitator™. The levitation force is then plotted as a function of the separation between the magnet and the levitator™. The zero-distance levitation force, F_o, is then estimated by extrapolating the curve to zero separation.

Figure 5 shows the zero-distance levitation force, F_o, for a batch of 39 hexagonally shaped samples. F_o is found to be distributed between 6 to 26 N (6.19 N to 25.93 N, to be specific). In two samples with F_o below 10 N, the seeds didn't initiate the grain growth during melt processing and resulted in a multi-domain structure. The variation of F_o between 11 to 26 N may result from several factors:

(i) improper seed selection: Although care has been taken to select single crystal seeds, some are polycrystals (two or three single crystals embedded with slight misorientation) which result in two or three domains. The levitation forces of these samples are generally low.

(ii) edge nucleation: Another problem observed during processing of levitators™ with large batch sizes is nucleation at the edges of hexagons. The levitation force of these levitators™ are also lower than that of single domains. Edge nucleation can be minimized by applying a thin layer of Yb_2O_3 at edges.

(iii) misorienation within a domain: Due to variations during the growth condition, small misoriented regions or sub-domains form in a domain. These c-axis misalignments between adjacent sub-domains are usually below 5 degrees. Two types of sub-domain structure are generally observed in single domain levitators™. A blocky sub-domain structure (~ 1 mm^2) is observed in the c-axis growth region whereas a strip-like sub-

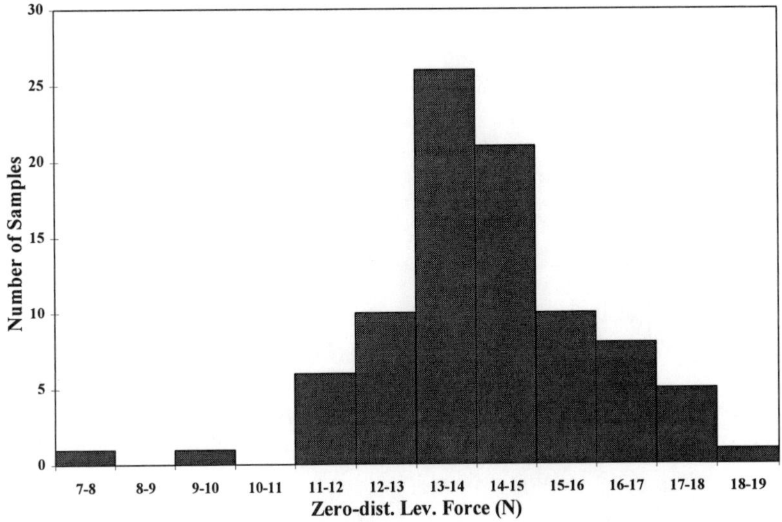

Figure 6: Distribution of Zero-distance levitation force for a batch of 90 levitators™.

domain structure (~ 0.1 mm^2) is observed in a/b axis growth region. These sub-domains may result in a substantial misorientation (as high as 15 degree) in the overall sample (from center to the sides) and can impact the levitation force of the samples. Work is now in progress to correlate this misorientation with the levitation force of a levitator™.

Figure 6 show the distribution of F_o for a batch size of 90 levitators™. The levitation force is found to vary between 7 to 19 N. Again the seeds were diffused during melt processing for two samples and resulted in a multi-domain structure. The F_o for these samples were below 10 N. The variation in the levitation force between 10-19 N may resulted from the factors discussed earlier

The results of Figures 5 and 6 show that single domain YBCO levitators™ can be fabricated in large batch sizes. Although, levitation forces (> 12 N) are sufficient for various practical purposes, further optimization is necessary to obtain results similar to that of smaller batch sizes.

CONCLUSION

Large domain YBCO levitators™ were fabricated in large batch sizes by melt processing techniques using $Nd_{1+x}Ba_{2-x}Cu_3O_y$ seeds to initiate the grain growth. A distribution in F_o is observed within a batch. The results from processing large batch sizes indicate that industrial-scale manufacturing of these levitators™ with F_o sufficient for frictionless bearing application is feasible.

ACKNOWLEDGMENTS

We would like to thank Dr. John Hull at Argonne National Laboratory for measuring the levitation forces.

REFERENCES

[1]F. C. Moon, *Superconducting Levitation, Applications to Bearings and Magnetic Transportation*, 1st ed., 198-212, (1994).

[2]J. R. Hull, T. M. Mulcahy, K. L. Uherka, and R. G. Abboud, " Low Rotational Drag in High-Temperature Superconducting Bearings," Proc. Applied Superconductivity Conf. Boston, Oct 1994.

[3]M. Morita, L. Trouilleux, S. Takebayashi, K. Kimura, M. Tanaka, K. Miyamoto and M. Hashimoto, "Growth of QMG Crystals," Proc. Int. Workshop on Superconductivity, Honululu, pp. 119-122, (1992).

[4]K. Y. Blohowiak, D. F. Garrigus, T. S. Luhman, K. E. McCrary, M. Strasik, I. A. Aksay, F. Dogan, W. B. Hicks, J. Liu, and M. Sarihaya, " Evaluation of $YBa_2Cu_3O_{7-x}$ Bulk Superconductors For High Field Magnet Applications," IEEE Trans. *Appl. Supercond.* **3**, pp. 1049-1052 (1993).

[5]M. Murakami, T. Oyama, H. Fujimoto, T. Taguchi, S. Gotoh, Y. Shiohara, N. Koshizuka, and S. Tanaka, "Large Levitation Force due to Flux Pinning in YBCO Superconductors Fabricated by Melt-Powder-Melt-Growth Process," *Jpn. J. Appl. Phys.* **29**, pp. L1991-L1994 (1990).

[6]R. Weinstein, I. -G. Chen, J. Liu, D. Parks, V. Selvamanikam, and K. Salama, "Persistant Magnetic Fields Trapped in High T_c Superconductor, " *Appl. Phys. Lett.* **56**, pp. 1475-1477, (1990).

[7]V. R. Todt, S. Sengupta, D. Shi, J. R. Hull, P. R. Sahm, P. J. McGinn, R. B. Poeppel, " Processing of Large $YBa_2Cu_3O_x$ - Domains for Levitation Applications by $NdBa_2Cu_3O_x$-Seeded Melt-Growth Technique," *J. of Electronic Materials* **23**, pp. 1127-1130 (1994).

[8]U. Balachandran, W. Zhong, C. A. Youngdahl and R. B. Poeppel, " Microstructure and Associated Properties of YBCO Superconductors Prepared by Melt-Processing Techniques," *J. of Electronic Materials* **22**, pp. 1285-1288, (1993).

[9] S. Sengupta, J. Corpus, J. R. Gaines, Jr., V. R. Todt, X. F. Zhang, D. J. Miller, C. Varanasi, and P. J. McGinn, "Fabrication and Characterization of Melt-

Processed YBCO", to be published in Proc. Of Applied Superconductivity Conference, Pittsburgh (1996).

[10]W. Lo, D. A. Cardwall, C. D. Dewhurst, and S-L. Dung, "Fabrication of large grain YBCO by seeded peritectic solidification", *J. Mat Res.* **11**, pp. 786-794 (1996).

[11]U. Balachandran, R. B. Poeppel, J. E. Emerson, S. A. Johnson, M. T. Lanagan, C. A. Youngdahl, D. Shi, K. C. Goretta, and N. G. Eror, " Synthesis of Phase-Pure Orthorhombic $YBa_2Cu_3O_x$ under Low Oxygen Pressure," *Materials Lett.,* **8**, pp. 454-456 (1989).

EFFECT OF Y_2BaCuO_5 PARTICLE SIZE ON UNDERCOOLING AND SOLIDIFICATION RATE OF $YBa_2Cu_3O_{7-x}$

F. Dogan and J. D. Reding

Department of Materials Science and Engineering,
University of Washington, Roberts Hall, Box 352120
Seattle, WA, 98195

ABSTRACT

The solidification rate of $YBa_2Cu_3O_{7-x}$ (123) was investigated in dependency of Y_2BaCuO_5 (211) particle size in the liquid phase. Microstructural development and reaction sequences of platinum doped and undoped 123 samples were studied by microscopy and Differential Thermal Analysis (DTA) techniques. Experimental studies revealed that solidification of 123 phase occurs at a more rapid rate and smaller undercooling (ΔT) as the size of 211 particles in the melt decreases due to the effect of platinum addition. The results suggests that the critical nucleation and growth temperature of 123 within the melt is controlled by the degree of undercooling which, in turn, depends on the size or dissolution rate of 211 particles.

I. INTRODUCTION

To achieve high critical current densities (J_c) in $YBa_2Cu_3O_{7-x}$ (123) superconductors, melt texturing techniques are widely accepted to minimize grain boundary weak links unavoidable in polycrystalline materials [1]. Microstructural development during melt texturing results in grain alignment and highly oriented texture. Other microstructural futures such as Y_2BaCuO_5 (211) inclusions within 123 matrix also contributes to the enhancement of J_c. In this aspect, fine dispersion of 211 inclusions are attained by Pt addition which effectively hinders the growth of 211 particles in the melt above the peritectic decomposition temperature [2]. A better control of the growth rate and microstructure of 123 phase can be achieved if the role of 211 size during solidification reaction is well understood.

Melt processed 123 samples were obtained by continual cooling as well as constant undercooling methods, and the similarity between these two methods were discussed with respect to solidification kinetics in [3]. The relationship between the growth rate and undercooling was reported in [4]. It is widely accepted that the peritectic reaction of 123 formation occurs by dissolution of 211 particles in the melt which in turn increases the Y solute concentration in the liquid [4-7]. The rate-limiting factor for the 123 growth rate taken as the dissolution rate of 211 phase which is controlled by the undercooling and the size of 211 particles.

In order to study the effect of 211 size on the solidification rate of 123, we used undoped and Pt added powders by continual cooling method. The solidification behavior of both samples were analyzed by Differential Thermal Analysis (DTA) method in conjunction with the microstructural development of quenched and melt textured samples. The relationship between 211 particle size in the melt, undercooling and solidification rate of 123 was discussed.

II. EXPERIMENTAL METHODS

Undoped and 0.5 wt.% Pt added 123 powders were provided by Praxair Specialty Ceramics, Seattle, WA. The powders were phase pure as determined by X-Ray Diffraction (XRD) analysis. The average particle was 2 μm as analyzed by Microtrac - X100 Particle Size Analyzer. The solidification behavior of semisolid phase was observed by DTA. The samples were heated at 10°C/min to 1100°C, soaked for one hour, and cooled at 1°C/min to 800°C. Pellets (10 mmØ and 5 mm thickness) were prepared by uniaxial pressing with subsequent isostatic compaction at 100MPa. Melt texturing of the samples were conducted on dense polycrystalline magnesia substrates.

To study the microstructural development of semisolid melt, the samples were heated to 1040°C with 2°C/min, held for 5 hours, and quenched in the slightly opened furnace to room temperature. The cooling rate was sufficient to prevent nucleation and growth of 123 phase. Melt textured grown (MTG) samples were prepared with the same heating profile and cooled at 2°C/hour to 950°C.

The samples were mechanically polished, etched with 1% dilute hydrochloric acid for a few seconds, followed by washing with alcohol. Microstructural characterization of samples was conducted by SEM JEOL 840A with EDS attachment.

III. RESULTS AND DISCUSSION

Figure 1 shows the SEM micrographs exhibiting the morphology and size of 211 particles of (a) undoped and (b) Pt-added 123 samples quenched after heat treatment at 1040°C. Comparing both microstructural images, we observe the formation of relatively large and bulky 211 particles in the melt of undoped 123. The shape of 211 particles varies from round with no apparent crystal habits on the surface to pillar with fat and short rod shape. Pt-doped samples, on the other hand, reveal the growth of fine acicular particles in the liquid with a high aspect ratio of length/width. A higher dissolution rate of 211 in the liquid may be attributed to smaller size or a larger total surface area of 211 particles.

It should also be mentioned that no liquid loss in Pt-doped 123 takes place during melt texturing process while depletion of liquid phase may be significant in undoped 123 depending on the type of substrate. We suggest that the retention of the liquid phase may be attributed to strong capillary forces between the fine 211 particles and the liquid phase [8]. Acicular shape of 211 particles, forming a

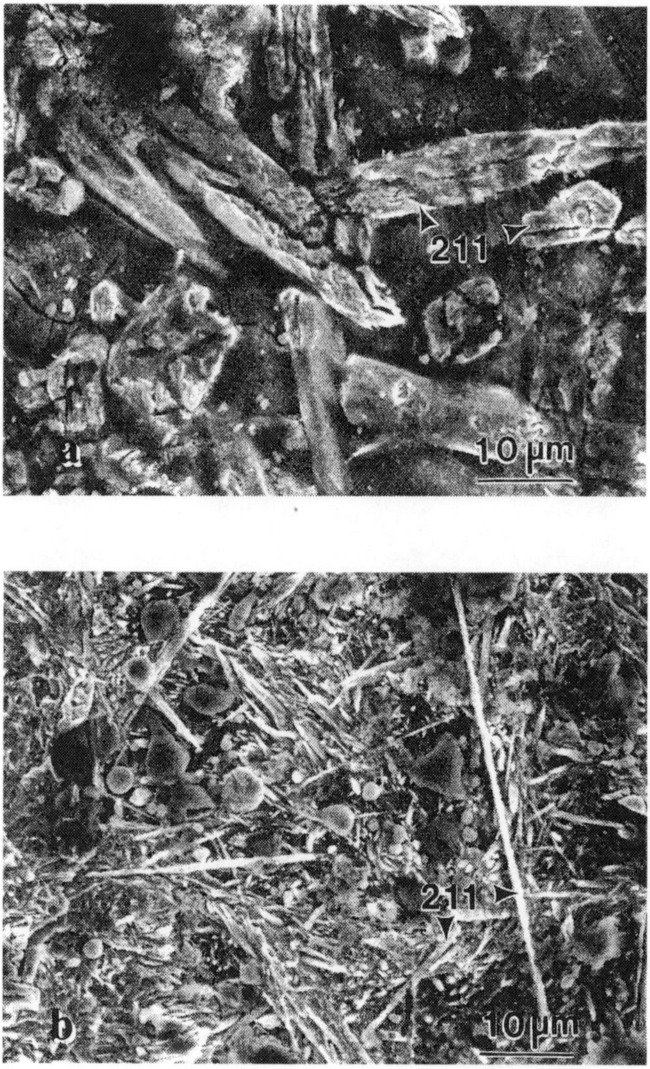

Figure 1 SEM micrographs of quenched (a) undoped and (b) platinum doped 123 samples, showing the size and morphology of 211 particles in liquid.

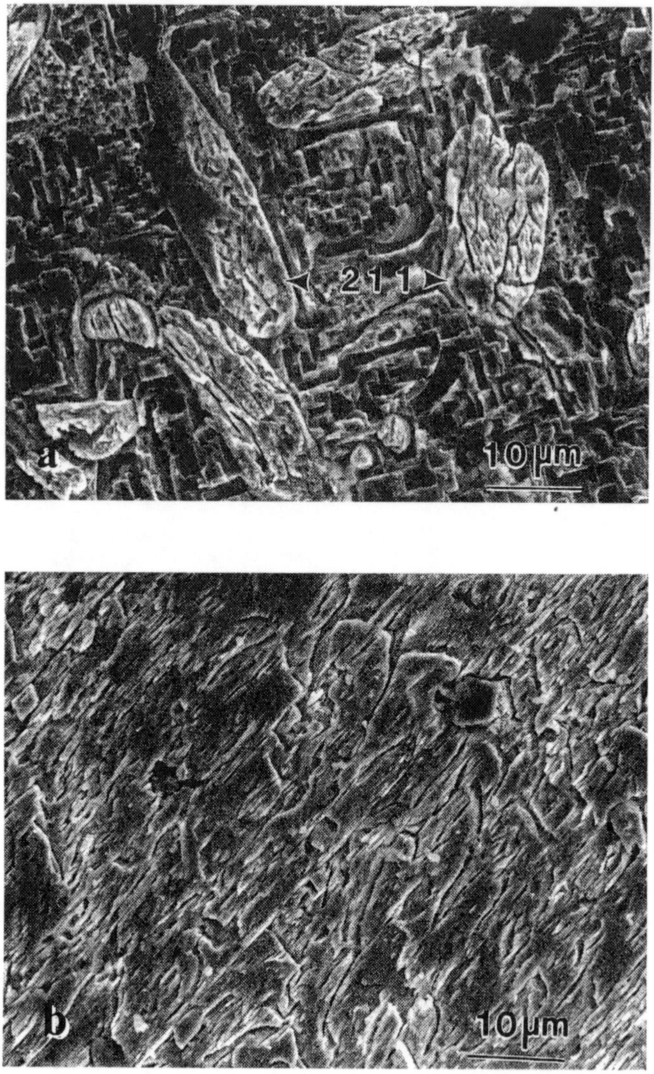

Figure 2 SEM micrographs of melt textured (a) undoped and (b) platinum doped samples. The undoped sample contains large 211 inclusions, while the platinum doped sample is 123 without 211 particles.

skeleton framework, was also considered to be effective for minimizing a liquid loss [9].

Shown in Fig. 2 are the micrographs of melt textured grown (a) undoped, and (b) Pt-doped samples. The microstructure of the undoped sample reveals 211 inclusions trapped within 123 matrix. The formation of excess 211 phase is due to depletion of the liquid phase above the peritectic temperature. The shape of trapped 211 particles are round and elongated which is similar to that as observed in quenched samples. On the other hand, Pt-doped sample reveals a microstructure free of 211 inclusions. This indicates that a stoichiometric 123 phase solidifies due to the complete retention of liquid phase in the semisolid melt.

Coarsening rate of the 211 phase was investigated by several research groups and explained by the modified Ostwald ripening theory [10-12]. Understanding the exact mechanism for refinement of 211 particles is very important to obtain fine dispersion of nonsuperconducting inclusions, resulting in the achievement of high Jc. It is generally accepted that the growth rate of 123 increase with decreasing size of 211 particles. However, a thorough understanding of the relationship between undercooling and the dissolution rate of 211 phase is necessary to control the growth rate of 123. This relationship becomes particularly important in growing large single grain 123 by seeded melt texturing technique. Therefore, in this work, the reaction sequences during solidification of 123 was monitored by DTA to gain insight into the interrelation between 211 dissolution rate and undercooling (ΔT).

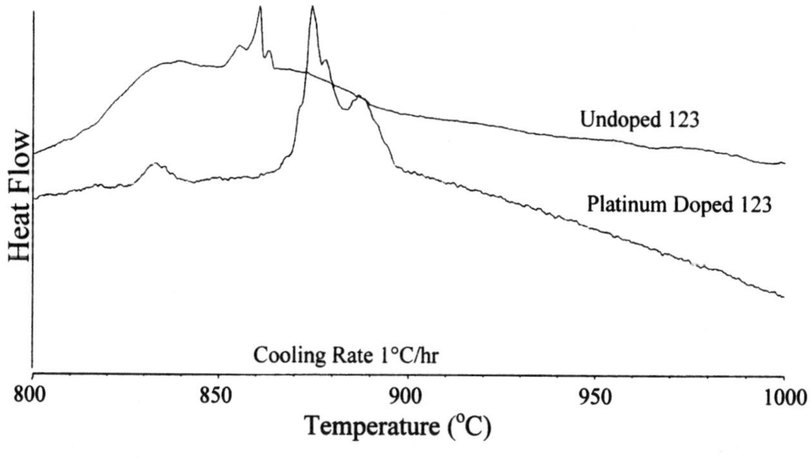

Figure 3 DTA plot of Pt doped and undoped 123 samples during cooling of semisolid phase.

Fig. 3 shows DTA plots during cooling of the semisolid undoped and Pt-doped samples. The onset of solidification temperature is at around 900°C for Pt-doped sample while slightly lower for undoped 123. It can also be depicted that a rather rapid dissolution of 211 particles takes place at a small undercooling which in turn corresponds to a higher solidification rate of 123 phase. On the other hand, a sluggish dissolution reaction of 211 or formation of 123 are observed during cooling of undoped 123 samples. Therefore, we conclude that the rate-limiting factor during solidification of 123 is the total surface area of 211 particles in the liquid which, in turn, corresponds to higher dissolution rate. Rapid dissolution of 211 phase is attributed to the higher surface free energy of fine particles [4].

While platinum acts as grain growth inhibitor for 211 particles, one can argue whether the growth kinetics of 123 phase are also affected by the presence of Pt dissolved in liquid. The DTA peaks in Fig. 3, indicating possible endothermic and exothermic reactions, are superimposed. It is rather difficult to distinguish between these reactions whether they corresponds to dissolution of 211 or formation of 123 phases. Multiple peaks may also be attributed to broad particle size distribution of 211 phase. This can be explained by the surface curvature effects associated with a differential dissolution rate of larger and smaller 211 particles.

The effect of surface curvature on the dissolution rate, and consequently the change in the yttrium concentration near a spherical particle, was addressed in detail by Izumi et. al. [14] considering the Gibbs-Thompson equation, $\Delta T_r = 2\Gamma/r$, where Γ is the Gibbs-Thompson coefficient, ΔT_r is the undercooling, and r is the radius of the particle. Due to the change in chemical potential, less supercooling will be required to drive the dissolution of the Y211, allowing the reaction to start at a higher temperature.

Although detailed studies have been conducted on growth and dissolution phenomena of 211 particles in Pt-doped liquid, further work remains to understand the role of platinum with respect to the growth rate of 123 phase. These studies may be very useful for a proper selection of additives which promote the growth of 123 phase during melt texturing while inhibiting the growth of 211 particles in liquid.

IV. SUMMARY

The effect of Y_2BaCuO_5 (211) particle size on the solidification rate of $YBa_2Cu_3O_{7-x}$ (123) phase during melt texturing process has been studied experimentally by Differential Thermal Analysis (DTA) technique. The results were discussed by comparing the microstructural development of Pt-doped and undoped samples. It has been shown that a small undercooling (ΔT) is sufficient to complete solidification reactions below the peritectic temperature in Pt-doped sample as the size of 211 particles decreases. The results indicate that the nucleation as well as the growth rate of 123 are controlled by an interplay between the extend of undercooling and the particle size of 211 phase in liquid. Although detailed studies have been conducted related to the role of dopants, such as Pt, on the growth kinetics of 211 particles, further investigations will be necessary to understand the effect of the additives on the growth rate of 123 phase.

V. ACKNOWLEDGMENTS

This work was supported by the Washington Technology Center under Contract # 95-A23. The Thermal Analysis equipment was available through a grant from NSF and NETZSCH, grant # DMR9410981.

VI. REFERENCES

[1]S. Jin, T. H. Tiefel, R. C. Sherwood, R. B. van Dover, M. E. Davis, G. W. Kammlott, and R. A. Fastnacht, "Melt Textured Growth of Polycrystalline $YBa_2Cu_3O_{7-x}$ with High Transport J_c at 77K," *Phys. Rev. B* **37**, 7850-3 (1988).

[2]N. Ogawa, I. Hirabayashi, and S. Tanaka, "Preparation of a high-J_c YBCO Bulk Superconductor by the Platinum Doped Melt Growth Method," *Physica C* **177**, 101-5 (1991).

[3]A. Endo, H. S. Chauhan, Y. Nakamura, and Y. Shiohara, "Relationship between Growth Rate and Undercooling in Pt-added $YBa_2Cu_3O_{7-x}$," *J. Mater. Res.* **11**, 1114-19 (1996).

[4]M. J. Cima, M. C. Flemings, A. M. Figueredo, M. Nakade, H. Ishii, H. D. Brody, and J. S. Haggerty, "Semisolid Solidification of High Temperature Superconducting Oxides," *J. Appl. Phys.* **72**, 179-90 (1992).

[5]T. Izumi and Y. Shiohara, "Growth Mechanism of $YBa_2Cu_3O_y$ Superconductors Prepared by the Horizontal Bridgman Method," *J. Mater. Res.* **7**, 16-23 (1992).

[6]C. A. Bateman, L. Zhang, H. M. Chan, and M. P. Harmer, "Mechanism for the Peritectic Reaction and Growth of Aligned Grains in $YBa_2Cu_3O_{6+x}$," *J. Am. Ceram. Soc.* **75**, 1281-3 (1992).

[7]A. Goyal, K. B. Alexander, D. M. Kroeger, P. D. Funkenbush, and S. J. Burns, "Solidification of $YBa_2Cu_3O_{7-x}$ from the Melt," *Physica C* **210**, 197-212 (1993).

[8]F. Dogan, "Growth of Stoichiometric $YBa_2Cu_3O_x$ Large Single Crystals without Y_2BaCuO_5 by Melt Texturing Technique," in *Program and Extended Abstracts for 1997 International Workshop on Superconductivity Co-Sponsored by ISTEC and MRS*, 205-7 (1997).

[9]N. Sakai, S. I. Yoo, and M. Murakami, "Control of Y_2BaCuO_5 Size and Morphology in Melt-Processed $YBa_2Cu_3O_{7-x}$ Superconductor," *J. Mater. Res.* **10**, 1611-21 (1995).

[10]T. Izumi, Y. Nakamura, and Y. Shiohara, "Doping Effects on Coarsening of Y_2BaCuO_5 Phase in Liquid," *J. Mater. Res.* **8**, 1240-6 (1993).

[11]C. Varanasi and P. J. McGinn, "Y_2BaCuO_5 Particle Coarsening during Melt Processing of $YBa_2Cu_3O_{7-x}$," *J. Electron. Mater.* **22**, 1251-7 (1993).

[12]M. L. Griffith, R. T. Huffman, and J. W. Holloran, "Formation and Coarsening Behavior of Y_2BaCuO_5 from Peritectic Decomposition of $YBa_2Cu_3O_{7-x}$," *J. Mater. Res.* **9**, 1633-42 (1994).

[13]S. P. Arthur, V. Selvamanickam, U. Balachandran, and K. Salama, "Study on Growth Kinetics in Melt-Textured $YBa_2Cu_3O_{7-x}$," *J. Mater. Res.* **11**, 1976-89 (1996).

[14]T. Izumi, Y. Nakamura, and Y. Shiohara, "Diffusion Solidification Model on Y-System Superconductors," *J. Mater. Res.* **7**, 1621-8, (1992).

A GROWTH MODEL FOR SEEDED MELT PROCESSING OF SINGLE-DOMAIN $YBa_2Cu_3O_x$

Dehui Qu, Brian Tent, Donglu Shi
Department of Materials Science and Engineering
University of Cincinnati
Cincinnati, OH 45221-0012

ABSTRACT

A growth model is proposed in the study of the growth mechanism of single-domain $YBa_2Cu_3O_x$ (Y123) using the seeded melt growth method. In this model, two important processes have been identified: one is the solidification which is treated as a simple "atom packing" process at a proper undercooling; the other is the diffusion of yttrium in the liquid from Y_2BaCuO_5 (Y211) to the growth front of solidification. Using this model, phenomena such as the dendrite growth pattern at the growth front and the step feature on the top surface have been explained.

INTRODUCTION

It has been reported that, using the Seeded-Melt-Growth (SMG) method, large single-domain superconducting materials such as $YBa_2Cu_3O_x$ (Y123) have been successfully produced and have shown a great potential for applications such as the flywheel energy storage device, microwave communication systems, and other electronic applications [1,2,3]. Investigations of the growth mechanisms of single-domain Y123 material have been conducted in this study in order to improve the performance of the material [4,5].

It has been commonly identified that the growth pattern on the top surface of single-domain Y123 shows a rectangular growth front between the solid and liquid phases and that growth originates in the seed and proceeds all the way to the edge of the sample. During the growth period, undercooling and yttrium diffusion in the liquid play the most important roles. Undercooling provides a driving force for the solidification while diffusion is responsible for carrying the yttrium cation from Y_2BaCuO_5(Y211) particles to the growth front. It is already known that the yttrium's solubility and diffusivity are very low in the liquid, leading to a slow growth rate at the processing temperature. Many investigations have been carried out to study the growth patterns, dendrite growth and undercooling control [4,5,6].

In this paper, the investigation will focus on the growth of single-domain Y123 using the SMG method. The growth mechanism will also be modeled based upon the results of the investigation.

EXPERIMENT AND DISCUSSION

Processing of the Material and Results

The Seeded-Melt-Growth (SMG) method has been developed to make large single-domainY123 samples with high critical current densities and high levitation forces. In our experiment, the commercially prepared Y123 and Y211 powders were used. After mixing in a proper ratio of Y123 and Y211(3:1) with a small amount of PtO_2 dopant (0.1~0.5 % wt), the powders were dry milled for approximately 12 - 24 hours. The well-mixed powders were then pressed into pellets uniaxially using a cylindrical or hexagonal die. $NdBa_2Cu_3O_x$ (Nd123) seeds were then selected and positioned on the center of the top surface of the pellets. Using alumina plates as substrates, the pellets were placed in a cylindrical furnace which has a vertical temperature gradient in the *c*-direction of the seeds. The temperature was increased to 1050°C at a rate of about 100°C/hr. After a short holding period (1 - 3 hrs), the samples were slowly cooled to a temperature below the peritectic temperature (T_P =1015°C) at a rate of 1 °C/hr. Figure 1 gives the heating schedule of the SMG method and Figure 2 shows the photograph of a typical single-domain Y123 sample.

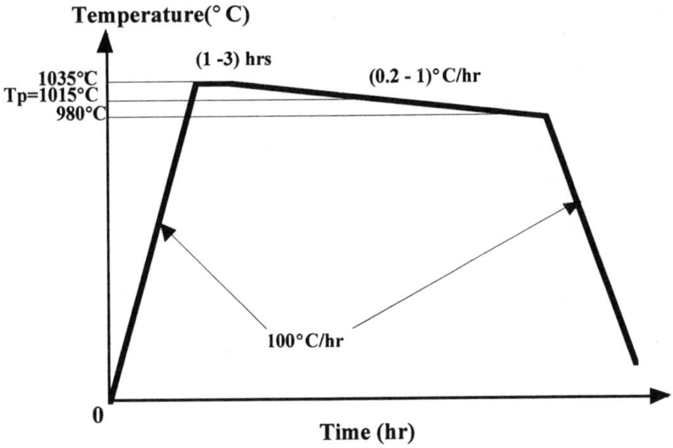

Figure 1 Heating schedule of the SMG method

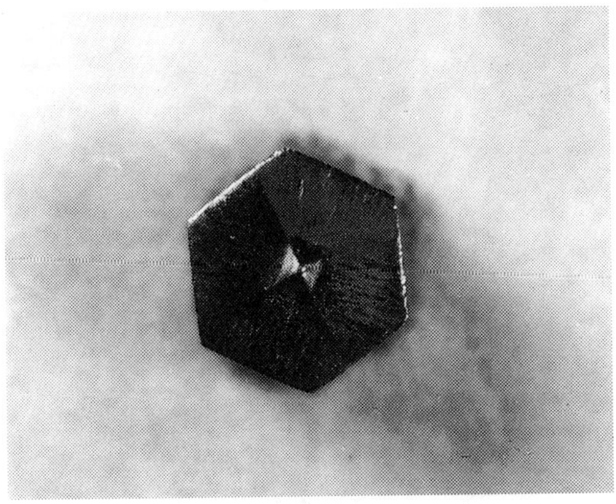

Figure 2 Single-domain $YBa_2Cu_3O_x$ samples made using the SMG method (22 mm)

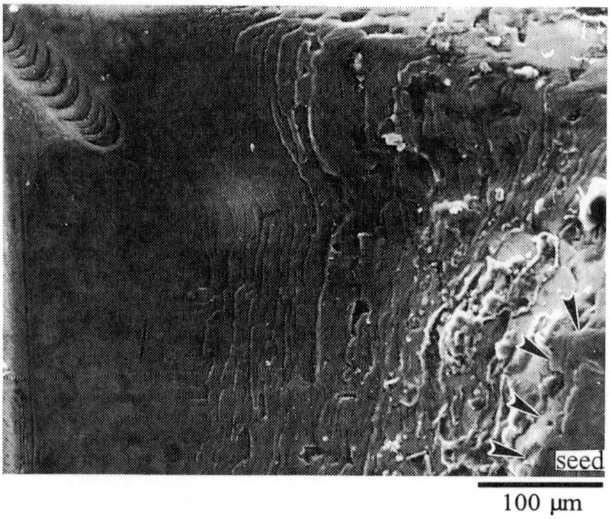

Figure 3a Low magnification (X 200) photograph of the area near seed

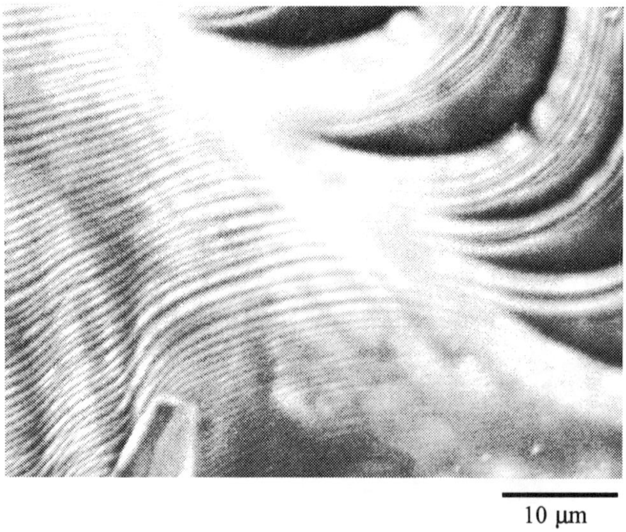

Figure 3b The step feature on the top suface at higher magnification (X 2000)

From the investigation of the top surface of the sample using SEM (Figure 3), it was noted that, although the seed used in this experiment is irregularly shaped, rectangular growth front lines were formed in each sample and the position of the seed grew outward from the center in a pyramid shape. The rectangular growth pattern on the top surface does not demonstrate a continuous growth process, but is rather formed by a series of rectangular facets featured with steps. The neighbored boundaries of these rectangular facets meet at the diagonal line of the rectangle extended from the seed and are connected by a smooth concave curve. It is also shown that the diagonal lines of the rectangular facets do not start exactly at the boundary between the seed and the previous liquid. It is observed that there is a small, flat plateau at the beginning of the rectangular growth. In the area close to the seed there are steps formed in an irregular shape similar to that of the seed. This could be thought of as a transition area between the seed and liquid.

It can be seen that the growth of Y123 from the peritectic reaction is not a continuous process. It is shown that there have been periodic stops throughout the whole growth. These periodic stops might be the explanation of the stepped features which result during the growth. A model explaining these features is suggested.

An "Atom Packing" Model

In order to further explore the reasons behind the periodic stops of the non-continuous growth process, an "Atom Packing" model is suggested. In this model, two important processes will be discussed, one is the solidification process which is treated as packing of moving atoms in the liquid onto the solid surface. The other is the diffusion process which is considered to be the transportation of the materials needed for packing which are not available on the spot.

Solidification is a process which is determined by the change of Gibb's free energy given by the following equation [7]:

$$\Delta G = \frac{L}{T_P} \bullet \Delta T$$

where L is the latent heat in the peritectic reaction, T_P is the temperature of peritectic reaction and ΔT is the undercooling of the solidification process (T_P-T). It can be seen from the equation that this is an undercooling controlled process in which the driving force (ΔG) is proportional to ΔT. The larger the undercooling is, the higher the driving force for this process and the faster this process will proceed.

Before solidification, atoms can move freely in the liquid. The seed used in the SMG method, which has the same crystal structure as Y123, provides a prototype of the Y123 crystal structure and yields the information about the correct positions of different atoms on the solid surface if solidification should happen. When there is a substantial undercooling in the local area of the seed, the atoms in the liquid will prefer to select the positions on the solid surface of the seed instead of in the liquid, due to the decrease of the Gibb's free energy. This atom packing process has to follow the stoichiometric composition of Y123. If any one of the components is consumed completely, the process will not continue.

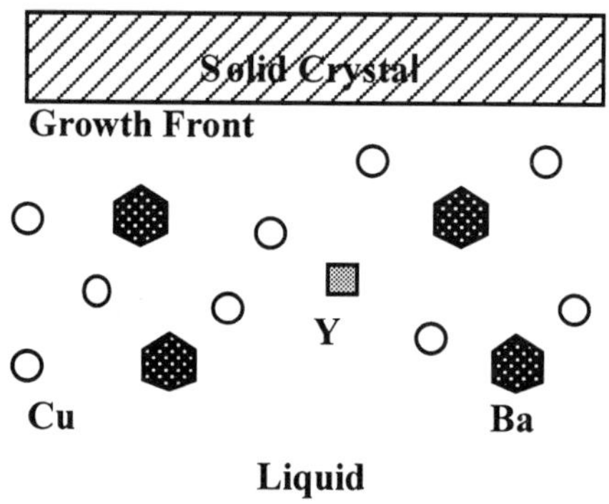

Figure 4 Schematic of atom movement in liquid before solidification

The diffusion process becomes very important when the atom packing process stops for the reason of being deficient of one component. It has been known that the liquid in which Y123 crystals grow is rich in the Ba and Cu cations, and the content of Y cation is low because of its small solubility in the liquid. The major source of the Y cation is the solid particles of Y211. In order to provide enough of the Y cation for the growth of the Y123 crystal, a diffusing process is necessary to transfer the yttrium from the Y211 particles to the solid growth front by diffusion which follows Fick's First law [7]:

$$J_Y = -D \bullet \frac{\partial C_Y}{\partial x}$$

where J_Y is the diffusing flow of yttrium from Y211 to Y123 growth front, D is the diffusion coefficient, and ($\partial C_Y/\partial x$) is the concentration gradient of yttrium between the growth front and the Y211 particle. At a certain value of the concentration gradient of yttrium, the diffusion coefficient D is the controlling factor of the process. If D is large enough, the growth will show a continuous process. If D is smaller, the packing process will be a non-continuous process and has a short waiting period due to the diffusion of yttrium. Once the yttrium cations are available, the process will continue and repeat periodically. So the model based on periodic non-continuous process can be used to explain the formation of the step feature on the top surface of a single-domain sample. But, if the value of D is extremely low, then the diffusion of yttrium will become the controlling factor for the entire growing process.

The diffusion of yttrium is not an independent process, but is influenced by undercooling. If the diffusion coefficient is treated as a constant in the range of the undercooling during this experiment, the flow of the yttrium cation from Y211 to the crystal growth front will increase slightly with the undercooling according to Fick's first law because faster packing processes will consume the yttrium cations quickly and make the concentration of yttrium at the growth front lower, so the concentration gradient of yttrium will increase.

From this model, it can be seen that, although the diffusion is treated as a stable process, the flow of yttrium cations is not a constant through the growing process. For a given undercooling, the concentrations of the yttrium cation can be thought of as constant at the growth front and the local area of Y211 particle and assumed to be a linear distribution. The concentration gradient of the yttrium cation will be a function of the distance between the Y211 particle and the growth front. So when the growth front gets close to a Y211 particle, the concentration gradient of yttrium will be larger and the flow of the yttrium will also increase. Then, in the local front area closest to this particle (indicated as zone A in the Figure 5), a higher yttrium concentration may be acquired, which will make the waiting period shorter in the solidification process. The growth rate will be faster in this area to develop a dendrite growth. The larger the undercooling is, the more easily the dendrite can form. This mechanism can be thought of as the major reason for the formation of dendrite morphology in the growth front of single crystal Y123. So it can be said that the large undercooling favors the formation of dendrites at the growth front while the small undercooling results in planar growth.

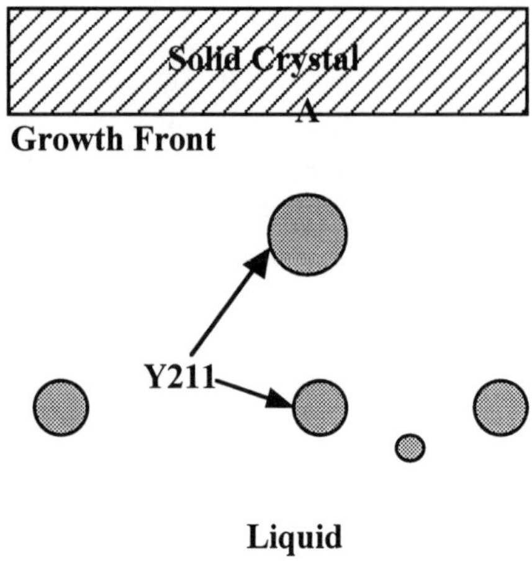

Figure 5 Schematic of influence of the distance between the growth front and Y211 particles to the yttrium flow

CONCLUSION

In the study of domain growth mechanisms of $YBa_2Cu_3O_x$ using the seeded melt growth method, a new "Atom Packing" model has been proposed. Combining the solidification and diffusion process, this model has revealed that periodic lapses exist during the growth and that the diffusing flow of yttrium is a function of the distance between the growth front and Y211 particles. Using this model, the phenomena during growth can be explained, such as the step feature of the top surface, planar growth pattern, undercooling control, and the dendrite morphology at the growth front of Y123.

REFERENCE:

1. D. Shi, W. Zhang, U. Welp, S. Sengupta, V. Todt, G.W. Crabtree, S. Dorris, and U. Balachandran, "Initial Crystallization and Growth in Melt Processing of

Large-Domain $YBa_2Cu_3O_x$ for Magnetic Levitation," IEEE Transaction on Magnetics, 5, 1627 (1994).

2. D. Shi, M. M. Fang, J. Akujieze, M. Xu, and J. G. Chen, "High Critical Current Density in Grain-Oriented Bulk $YBa_2Cu_3O_x$ Processed by Partial Melt Growth," Appl. Phys. Lett., 57, 2605(1990)

3. C. D. Dewhurst et al "Distribution of Critical Density in Large $YBa_2Cu_3O_{7-x}$ Grains Fabricated Using Seeded Peritectic Solidification" 1996 Applied Superconductivity Conference

4. D. Shi, K. Lahiri, S. Sagar, "A Domain Growth Pattern in Seeded-Melt Processing of $YBa_2Cu_3O_x$" High Temperature Superconductors: Synthesis, Processing, and Large-scale Applications. The Minerals, Metals & Materials Society, 1996

5. K. Salama, D. F. Lee, R. G. Presswood Jr., and T. Francavilla "Seeded Directional Solidification of $YBa_2Cu_3O_x$ Superconductors" Physica C 235-240(1994) 213-216

6. Y. Nakamura, A. Endo, Y. Shiohara, "The relation between the undercooling and the growth rate of $YBa_2Cu_3O_{6+x}$ superconductive oxide" J. Mater. Res., Vol. 11, No.5, May 1996

7. David A. Porter, Kenneth E. Easterling, "Phase Transformations in Metals and Alloys" (1990) Chapman and Hall

PRIMARY CRYSTALLIZATION VOLUME OF BSCCO 2212: APPLICATIONS TO CRYSTAL GROWTH AND MELT PROCESSING.

Lawrence P. Cook and Winnie Wong-Ng
Ceramics Division
NIST
Gaithersburg
MD 20899

ABSTRACT

Two crystallization paths have been studied in the Bi-Sr-Ca-Cu-O (BSCCO) system, based on the experimentally determined $Bi_2Sr_2CaCu_2O_x$ (2212) primary phase field. The crystallization experiments were completed in air, and the products were characterized using scanning electron microscopy (SEM) and energy dispersive x-ray analysis (EDS). Prior to crystallization, samples were melted at 1100 °C. The first crystallization path was chosen for a bulk composition lying just inside the high temperature part of the of the 2212 primary phase field. Slow cooling (0.5 °C/min) from 900 °C to 800 °C resulted in a microstructure characterized by long (>200 µm) laths of 2212, and lesser $Bi_{11}Sr_9Ca_xCu_5O_x$ (119x5) and CuO. The interstices between the laths were filled with a eutectic-like microstructure containing 2212, 119x5, CuO and $Bi_2Sr_1Ca_1O_x$ (2110). These features can be readily interpreted in terms of the 2212 primary crystallization field. The second crystallization path was chosen using a composition lying just outside the 2212 primary crystallization field in the $(Ca,Sr)_2CuO_3$ (0x21) primary field. Similar slow cooling resulted in a much different microstructure dominated by scattered large crystals of 0x21, as would be predicted from the primary crystallization field data. This particular bulk composition required partial resorption of the 0x21 during cooling in order to maintain equilibrium. However, the predicted path was not followed due to crystal flotation, which resulted in local equilibrium and produced sample inhomogeneities. By contrast, the first bulk composition produced a homogeneous distribution of large 2212 crystals relatively free of large second phase particles. These experiments emphasize the importance of primary phase field data in BSCCO superconductor melt processing.

INTRODUCTION

The development of superconducting wire and tape based on the Bi-Sr-Ca-Cu-O (BSCCO) 2212 phase (nominally $Bi_2Sr_2CaCu_2O_x$) has been pursued largely through melt processing methods [1-4]. Typically in this approach, precursor powder is packed into silver tubing, which is then drawn into wire or rolled into tape [5-8]. Ideally, the wire or tape then undergoes annealing at sufficiently high temperatures to produce partial melting, resulting in the desired microstructure and phase assemblage [9-12]. Multiple annealing and mechanical deformation steps may be employed to improve properties. In practice, the presence of unwanted second phases, usually alkaline earth cuprates, is thought to interfere with the processing. Optimization of the properties of the 2212-based melt-processed materials is at least partially dependent upon a solution to this problem.

The need for phase equilibrium data on the BSCCO superconductors has been generally recognized, and numerous partial studies have been published, of both 2212 subsolidus and liquidus phase relations [13-24]. Recently, we have completed a determination of the approximate 2212 primary crystallization field in air [24], as shown in Figure 1. This knowledge allows the design of processing paths for optimizing the content of 2212, as well as for controlling the microstructure and the second phase content. In the present paper, we examine two different 2212-based bulk compositions and cooling histories, with application to 2212 melt processing.

EXPERIMENTAL METHODS

Samples for the experiments were prepared using reagent grade powders of Bi_2O_3, $SrCO_3$, $CaCO_3$, and CuO. Mixtures were weighed corresponding to the compositions in Table I, which were chosen to be closely associated with the 2212 primary phase field. These mixtures were homogenized, pelletized, and calcined in air at 800 °C for 24 h, with repeated grindings. Samples of ~500 mg size were weighed into MgO crucibles, and placed into the differential thermal analysis/thermogravimetric analysis (DTA/TGA) apparatus. The apparatus was then evacuated and backfilled with purified (CO_2- and H_2O-scrubbed) air, which was flowed continuously through the system at a rate of ~10 cm^3/min during the experiments. The samples were held at 800 °C in the apparatus for several hours, until it was determined that there was no further weight loss. Next the samples were taken to 1100 °C and held for one hour. This step produced complete melting, as verified by independent quenching experiments at 1100 °C. After this step, samples were furnace-cooled to 900 °C to begin the crystallization experiments. Identical experiments were performed on the calcined compositions I and II of Table I, in which the melted samples were cooled at 0.5 °C from 900 °C to 800 °C, resulting in complete crystallization.

Table I. Nominal bulk compositions used in slow cooling experiments.

Sample	Composition (mole fraction, %)			
	$BiO_{1.5}$	SrO	CaO	CuO
I	26.9	24.7	17.6	30.8
II	28.7	21.4	20.8	29.1

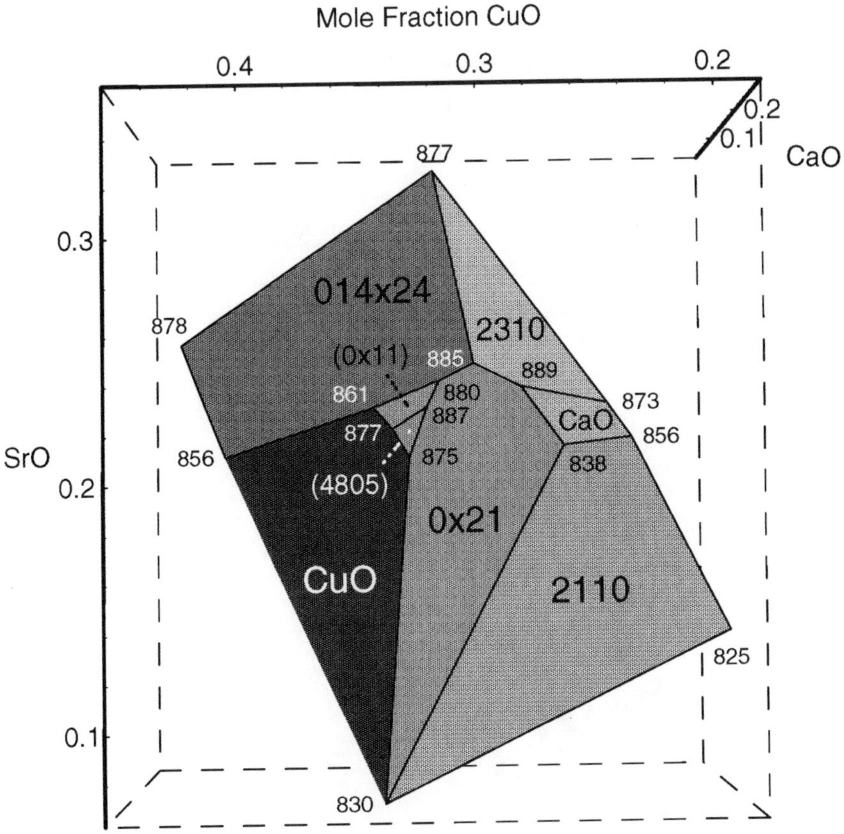

Figure 1. Primary crystallization field of 2212 in the BSCCO system [24].

The crystallized products were potted in epoxy, polished, carbon-coated, and examined by scanning electron microscopy (SEM), primarily using backscattered electron images, as well as by energy dispersive x-ray spectrometry (EDS). For the EDS measurements, we used an accelerating potential of 20.3 kV. A 2212-type superconductor with 4334 (Bi:Sr:Ca:Cu) stoichiometry was used as the analytical standard. Data collection and reduction were accomplished using conventional microanalytical methods [25] via the program DTSA [26]. Uncertainties are estimated as follows: for mole fractions > 10 %, < ± 5 % relative expanded uncertainty; for mole fractions < 10 %, < ± 10 % relative expanded uncertainty.

RESULTS AND DISCUSSION:

The results of slow cooling experiments were significantly different for compositions I and II of Table I. This would be expected, given the positions of these compositions relative to the 2212 primary phase field. Composition I plots inside the 2212 primary phase field, whereas composition II is nearby, but just outside the 2212 primary phase field.

Composition I:

The microstructure resulting from the slow crystallization experiment on composition I is shown in the micrograph of Figure 2. This microstructure is dominated by interlocking, relatively inclusion-free crystals of 2212. Similar crystals were observed in free-standing form on the surface of the sample, where the melt had apparently receded, leaving voids between the crystals. However, in the area of Figure 2, taken from deeper within the crucible, the interstices between 2212 crystals are filled with crystalline material. Grains of CuO and 119x5 (nominally $Bi_{11}Sr_9Ca_xCu_5O_x$) are present between the 2212 crystals, and the remaining interstices are filled with a mixture of 2212, 119x5, CuO and 2110 (nominally $Bi_2Sr_1Ca_1O_x$).

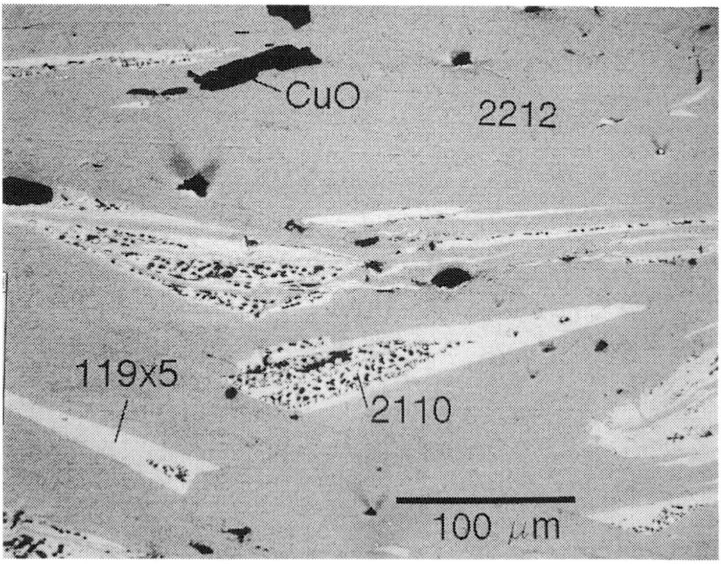

Figure 2. Backscattered electron micrograph of slow cooling experiment on bulk composition I.

Averaged EDS analyses of these phases are given in Table II, with the exception of the CuO, which contained no detectable cation impurities. The other phases are all solid solutions, and so their compositions would be expected to vary along the crystallization path. In general, this variation has been relatively small, however, as indicated by the standard deviations. Recalculated averages of Table II give estimated stoichiometries as follows for the 2212, 119x5 and 2110 phases, respectively: $Bi_{2.1}Sr_{1.8}Ca_{0.9}Cu_{2.0}O_x$, $Bi_{10.4}Sr_{7.4}Ca_{1.2}Cu_{5.0}O_x$, and $Bi_{2.0}Sr_{0.6}Ca_{1.2}Cu_{0.2}O_x$. The significance of the CuO in the 2110 is not clear. It is also present in 2110 produced during the composition II slow crystallization experiment (see Table III), and so it may indicate CuO solid solution in 2110, rather than the presence of finely distributed impurities.

Table II. Average phase compositions for slow cooling experiment, bulk composition I.

Phase	#Analyses	Composition (mole fraction, %)[1] $BiO_{1.5}$	SrO	CaO	CuO
2212	15	30.5 ± 1.4	26.8 ± 2.0	13.4 ± 0.5	29.3 ± 0.7
119x5	5	43.5 ± 1.0	30.7 ± 1.5	4.9 ± 0.7	20.9 ± 0.5
2110	4	50.5 ± 1.3	14.0 ± 3.3	29.4 ± 0.9	6.1 ± 3.0

[1] Stated uncertainties are one standard deviation, as derived from statistical analysis of multiple data points. Refer to text for discussion of overall relative expanded uncertainties.

On the basis of Figure 2, which is typical of the slow crystallization experiment on composition I, three microstructural components are present: A) large 2212 crystals of several mm length; B) medium-sized (20 µm - 100 µm) interstitial grains of CuO and 119x5; and C) fine-grained (generally < 5 µm) intergrowths of 2212, 119x5, CuO and 2110. Each of these microstructures can be interpreted in terms of the 2212 primary crystallization field data of Figure 1, as explained in more detail by reference to Figure 3, using well-known principles of multicomponent phase equilibria [27].

As indicated in Figure 3, composition I plots just inside the 2212 primary phase field. The initial part of the crystallization path is defined by a line passing from 2212 through bulk composition I. As the temperature of the composition I melt is cooled, it falls within the 2212 primary phase field and begins to crystallize, with the composition of the remaining liquid moving away from composition I along the line in a direction away from 2212. This part of the crystallization path is

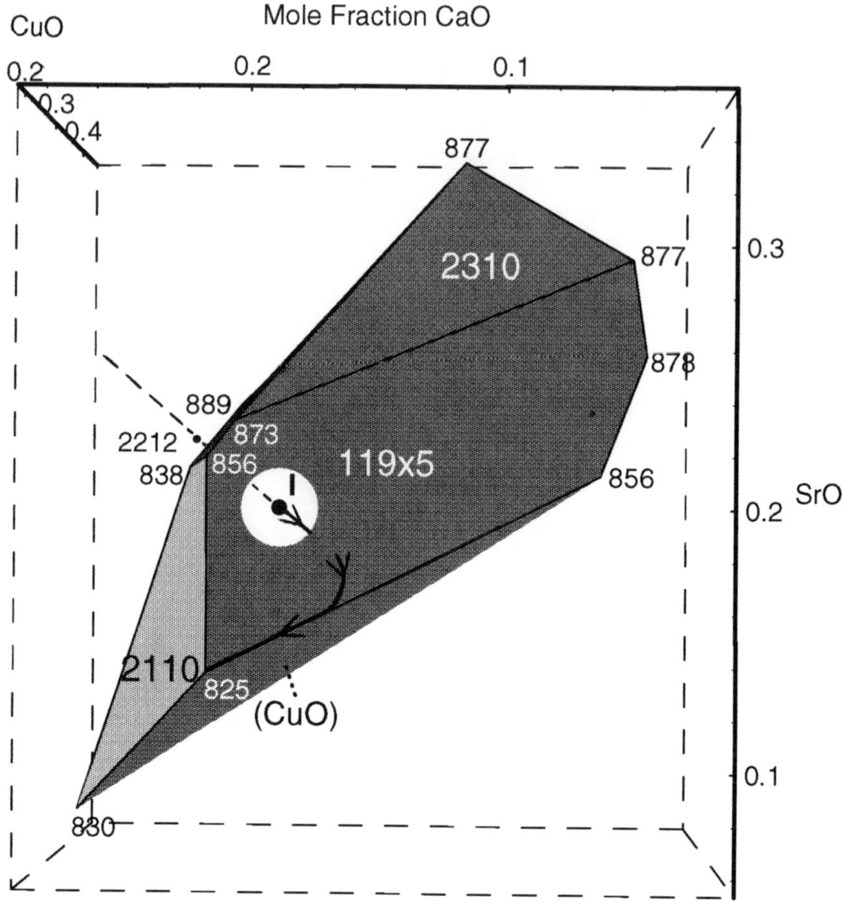

Figure 3. Suggested crystallization path for bulk composition I.

responsible for the large crystals of 2212 corresponding to the type A microstructure indicated above. With falling temperature, the liquid continues to move along the line until it intersects the outer limit of the 2212 primary phase field which is defined by the 119x5 cotectic surface. At this point, 119x5 begins to crystallize simultaneously with 2212. The crystallization path changes direction, and is defined by the intersection of the 119x5+2212 cotectic surface with a plane containing 2212, 119x5 and bulk composition I. Crystallization continues along this curve until the 2212+119x5+CuO cotectic line is intersected, which now becomes the crystallization path, and CuO crystallizes together with 2212 and

119x5. These parts of the crystallization path produced the medium-sized crystals of 119x5 and CuO noted above in the type B microstructure. Crystallization along the cotectic line continues until the 5-phase reaction point: 2212+119x5+CuO+ 2110+liquid is reached. This is the minimum melting point on the 2212 primary crystallization volume. Since it is a eutectic, all of the liquid crystallizes to produce a mixture of 2212, 119x5, CuO and 2110. This is the origin of the fine-grained 4-phase intergrowths of the type C microstructure noted above.

The proposed crystallization history can be checked by recalculating the bulk composition in terms of the final equilibrium assemblage. Levin et al. [27] have given an extensive discussion of this principle with regard to ternary systems, and it may easily be extrapolated to quarternary systems. The requirement is that, at equilibrium, the final crystallization product must be consistent with the subsolidus multiphase field within which the bulk composition plots. As Table IV shows, bulk composition I plots within the four phase field 2212+119x5+2110+CuO, using a reasonable choice of solid solutions based on the phase compositions of Table II. Therefore, the requirement appears to be met.

Composition II:

Although experiments on compositions I and II were identical, composition II showed evidence of crystal/liquid segregation, in contrast to composition I. Figure 4 is a micrograph from near the top of the experiment, where the microstructure is dominated by large 0x21 crystals. These crystals were also observed on the top surface of the experiment, where they could be differentiated by their slight bronze coloration, relative to the surrounding 2212 crystals. As can be seen in Figure 4, the spaces between the 0x21 crystals are filled by medium-sized grains of 2212, lesser CaO, and a small amount of 2110, largely in the remaining interstices. Figure 5 is a micrograph from somewhat deeper in the crucible, where the large 0x21 grains of Figure 4 are completely lacking, and the microstructure instead is dominated by large grains of 2212 with lesser 119x5 and CuO. The difference in micrographs can be attributed to rise of 0x21 crystals to the top of the experiment as they formed. The lightest cations in the BSCCO system are concentrated in the 0x21 phase, and it would be likely that density differences between this phase and the surrounding liquid would cause crystals to float to the top.

Averaged EDS analyses of the phases observed in the slowly crystallized experiment on composition II are given in Table III, with the exception of CuO, which contained no detectable impurities. The variations in alkaline earth composition for the 119x5 phase are somewhat larger than for Table II, as indicated by the standard deviations of the averages. The variation in Sr/(Sr+Ca) over 7 analyses ranged from Sr/(Sr+Ca) = 0.58 to Sr/(Sr+Ca) = 0.77. This increased variation may have resulted from the sampling of finer 119x5 grains, some of which formed at lower temperatures near the final stages of crystallization. It is known that Ca-rich members of the 119x5 solid solution are stable at lower temperatures. Similary, the 0x21 phase showed a range of Sr/(Sr+Ca) ratios, from Sr/(Sr+Ca) = 0.07 to Sr/(Sr+Ca) = 0.17, over 5 analyses. This variation is consistent with its complex history, as discussed below. Based on the recalculated averages of Table III, representative phase stoichiometries for the 0x21, 2212, CaO, 119x5 and 2110 phases, respectively, are estimated as follows:

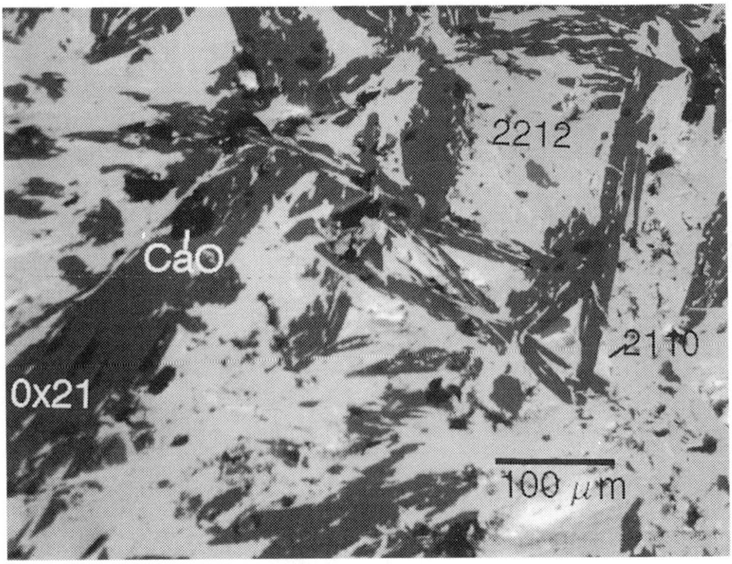

Figure 4. Backscattered electron micrograph of slow cooling experiment on bulk composition II, near top of experiment.

$Ca_{1.8}Sr_{0.2}CuO_3$, $Bi_{2.1}Sr_{1.9}Ca_{1.1}Cu_{2.0}O_x$, $Ca_{0.98}Sr_{0.02}O$, $Bi_{10.7}Sr_{6.2}Ca_{2.5}Cu_{5.0}O_x$, and $Bi_{2.0}Sr_{0.5}Ca_{1.3}Cu_{0.1}O_x$. These stoichiometries are in general similar to those in Table II, except as noted.

The microstructures of Figures 4 and 5, and their differences relative to each other and relative to Figure 2, can be interpreted via the 2212 primary phase field data (Figure 6). As indicated in Figure 6, bulk composition II may be assumed to lie slightly outside the 2212 primary crystallization volume, near the 0x21+2212 cotectic surface, within the 0x21 primary crystallization volume. The initial crystallization path is then defined by a line extending from 0x21 through bulk composition II.. As the melt cools, it falls within the 0x21 crystallization field, and crystallization of 0x21 begins, with the liquid moving along the line away from 0x21. This part of the crystallization path is responsible for the large crystals of 0x21 observed in Figure 4. Crystallization of 0x21 continues, with the liquid moving along the line until the 0x21+2212 cotectic crystallization surface is intersected. At this point 2212 begins to crystallize and the crystallization path follows a curve defined by the intersection of the plane containing 0x21, 2212 and bulk composition II with the 0x21+2212 crystallization surface. Crystallization of

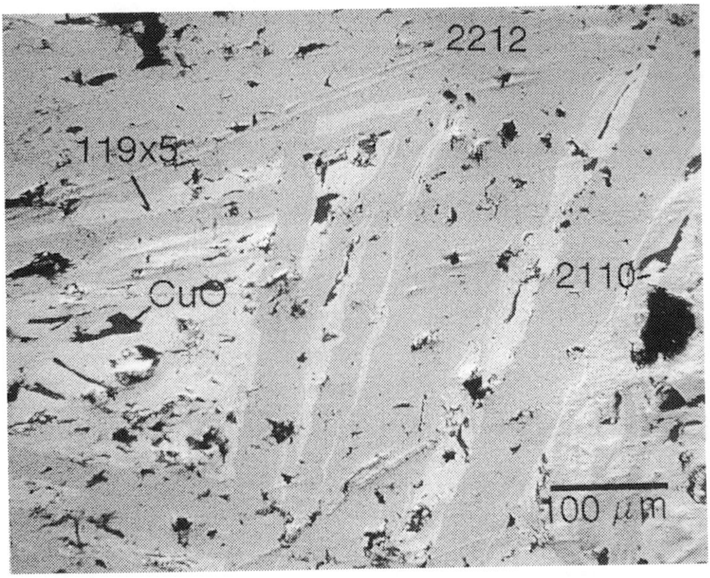

Figure 5. Backscattered electron micrograph of slow cooling experiment on bulk composition II, near interior of experiment.

2212 and 0x21 continue until the 0x21+2212+CaO cotectic line is intersected. At this point CaO begins to crystallize. Crystallization along the path continues until the 0x21+2212+CaO+2110 reaction point is reached. The microstructure of Figure 4 can be interpreted as having resulted from crystallization to this point. The large 0x21 crystals were the first crystals to form, and then crystallized together with the 2212 crystals, and finally with CaO and 2110. Note that the 0x21 crystals of Figure 4 have a ragged appearance. This suggests partial resorption, which is indeed necessary for the liquid to move along the 0x21+2212+CaO cotectic.

If equilibrium were maintained, the progression of the liquid for bulk composition II would continue from the 0x21+2212+CaO+2110 reaction point in Figure 6 once the CaO had been consumed by the reaction. The liquid would then move along the 0x21+2212+2110 cotectic line, with simultaneous crystallization of all three phases. Ultimately, the 0x21+2212+2110+CuO reaction point would be reached, and the final assemblage would consist of only these four phases, once the liquid had completely crystallized. This is required by the recalculation of Table IV, which indicates that bulk composition II plots within the subsolidus field defined by these four phases. However, it is obvious from Figure 4 that not all the CaO was consumed, CuO was not present, and the predicted path was not followed. Thus it

Table III. Average Phase Compositions for Slow Cooling Experiment, Bulk Composition II.

Phase	# Analyses	Composition (mole fraction, %)[1] $BiO_{1.5}$	SrO	CaO	CuO
2212	5	29.1 ± 1.1	26.9 ± 1.7	15.9 ± 0.8	28.2 ± 1.4
119x5	7	43.9 ± 0.6	25.5 ± 2.3	10.1 ± 1.7	20.5 ± 0.4
2110	7	51.2 ± 1.9	13.6 ± 3.2	32.7 ± 1.4	2.5 ± 0.5
0x21	5	0.0 ± 0.0	7.7 ± 2.5	60.1 ± 2.6	32.2 ± 0.3
CaO	5	0.0 ± 0.0	2.0 ± 0.1	96.9 ± 0.1	0.6 ± 0.1

[1] Stated uncertainties are one standard deviation, as derived from statistical analysis of multiple data points. Refer to text for discussion of overall relative expanded uncertainties.

Table IV. Sample bulk compositions recalculated in terms of final equilibrium four-phase assemblages.

Sample	Calculated phase composition (mole fraction, %) 2212	119x5	0x21	2110	CuO
I	92.3	1.2	---	2.9	3.6
II	69.0	---	7.2	16.9	6.

appears, that for this part of the sample, the crystallization path effectively stopped at the 2212+0x21+CaO+2110 equilibrium. This can be explained by a density-induced concentration of 0x21 crystals, as suggested above. This increase in 0x21 concentration effectively reduced the amount of liquid in the top part of the sample, making it difficult for the remaining liquid to chemically communicate with all the solid phases. For this reason, the crystallization path was terminated and did not continue to its predicted completion in this part of the sample. Also, there may have been a kinetic factor as well: since continued movement of the liquid along the equilibrium crystallization path moves it closer to 0x21, then 0x21 must be resorbed by the liquid for this to occur. It is likely that such dissolution of large grains into the liquid would be relatively slow, compared to crystallization.

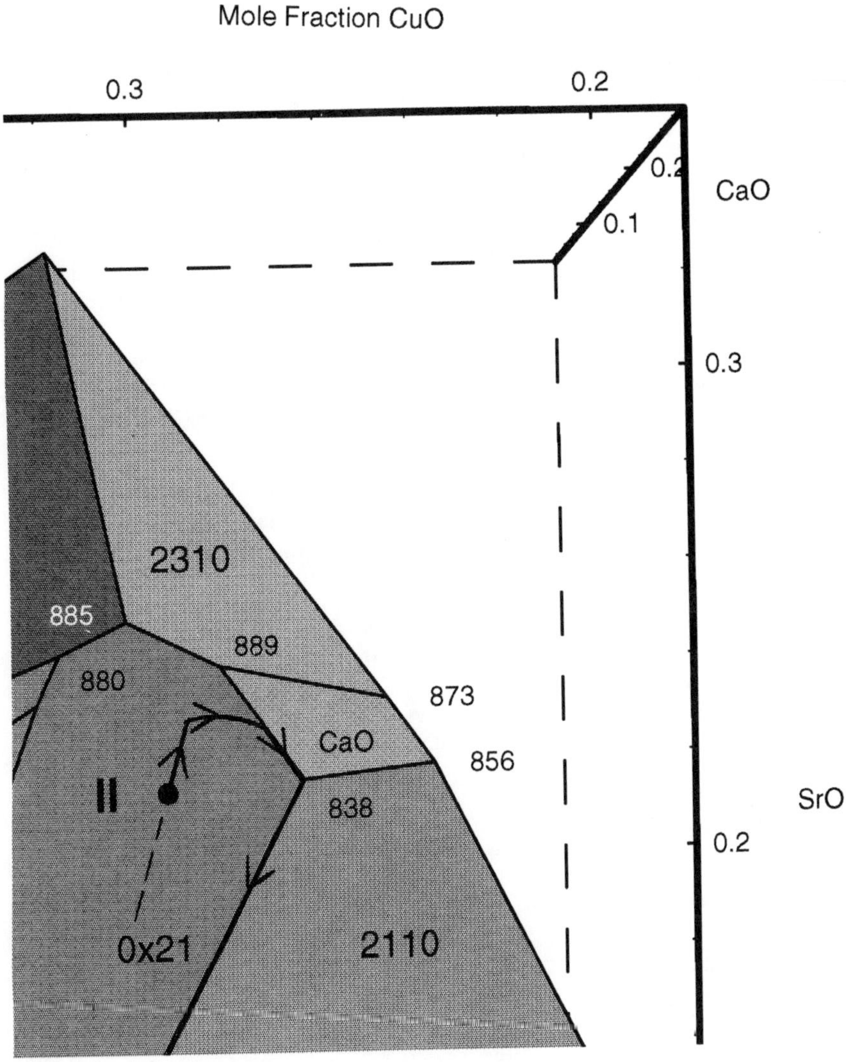

Figure 6. Suggested crystallization path for bulk composition II (enlargement of Fig. 1).

In the interior part of the sample, from which the 0x21 crystals floated away, the complementary effect was to increase the amount of liquid in that part of the sample. Furthermore, the bulk composition in that part of the sample was moved in a direction away from 0x21, placing it within the 2212 primary crystallization field. This explains the phase assemblage of Figure 5, i.e.:2212+119x5+CuO+2110, which corresponds to the final crystallization assemblage of Figure 3. The lack of a similar eutectic microstructure among the phases in Figure 5 is probably due to the absence of the relatively static crystallization situation which prevailed for Figure 2.

To summarize, for bulk composition II, crystal/liquid segregation effectively caused the sample to become separated into two new bulk compositions, so that the equilibrium crystallization requirements for bulk composition II could no longer be followed. However, under conditions of local equilibrium, it may be assumed that the requirements for equilibrium were met for each of the new bulk compositions, leading to different final phase assemblages in different parts of the sample.

CONCLUSION

Slow crystallization experiments on two bulk compositions closely associated with the 2212 primary crystallization field have demonstrated the importance of careful control of the cooling history and bulk composition during processing of the 2212 superconductor. For the composition chosen within the 2212 primary crystallization field, the microstructure was dominated by large interlocking 2212 crystals, with a relatively small amount of interstitial material. For a bulk composition with mole fractions a few % away, but which plots just outside the 2212 primary phase field, in the 0x21 primary phase field, the results were dramatically different, in that the microstructure was dominated by large crystals of 0x21. This was further complicated by flotation of 0x21 crystals, which produced local changes in the bulk composition. Also, the equilibrium resorption of 0x21 and CaO during cooling was kinetically difficult to maintain.

Clearly, for superconductor melt processing, it is essential to have bulk compositions within the superconductor primary phase field, as these experiments have demonstrated. Otherwise, particularly with respect to the alkaline earth cuprate phases, several effects, including not only primary equilibrium crystallization, but also local equilibrium and kinetic factors, are likely to result in large second phase particles, which interfere with the development of the desired superconductor microstructure. Such effects could be expected to have a major effect on bulk properties and performance.

ACKNOWLEDGEMENTS:

Partial support for this study was provided by the U. S. Department of Energy. W. Greenwood assisted with the sample preparation, and E. Hayward assisted with the illustrations.

REFERENCES

[1] Y.Shiohara, M. Nakgawa, T. Suga, K. Ishige, T. Oyma, T. Izumi, S. Nagaya, M. Miyajima, I. Hirbayshi, and s. Tanak, "Directional Solidification Processing of High T_c Superconducting Oxides," Advances in Supercnductivity II, Proceedings of the 2nd International symposium on superconductivity (ISS '89), Nov. 14-17, 1989, Tsukuba, Japan, T. Ishiguro and K. Kajimura, Eds., Springer-Verlag, 263-268.

[2] W. Zhang and E.E. Hellstrom, "The Development of an Aligned Microstructure during Melt Processing of $Bi_2Sr_2CaCu_2O_x$ (BSCCO-2212) Doctor-Bladed Films," *Physica C* , **218** 141-132 (1993).

[3] R. Zhou, W.L. Hults, J.F. Bingert, J.Y. Coulter, E.J. Peterson, and J.L. Smith, "Ag Ribbons in Bi2212 Powder-in-tube Tape," *Physica C* , **249** (1995) 166-170.

[4] T.D. Aksenova, P.V. Bratukhin, S.V. Shavkin and V.L. Melnikov, E.V. Antipova, N.E. Khlebova and A.K. Shikov, "Texture Formation in $Bi_2Sr_2CaCu_2O_x$/Ag Tapes Prepared by Partial Melt Process," *Physica C,* **205** 271-279 (1993).

[5] K.H. Sandhage, G.N. Riley, Jr., and W. Carter, "Critical Issues in the OPIT Processing of High-J_c BSCCO Superconductors," *J. Metals,* **43** 21 (1991).

[6] U. Balachandran and A.N. Iyer, P. Haldar, J.G. Hoehn, Jr. and L.R. Motowidlo, "Progress in Development of Tapes and Magnets Made From Bi-2223 Superconductors," *Proceedings of World Congress on Superconductivity*, (Orlando, FL. June 27-July 1, 1994).

[7] K. Sato, T. Hikata, H. Mukai, M. Ueyama, N. shibata, T. Kato, T. Masuda, M. Nagata, K. Iwata and T. Mitsui, *IEEE Trans Mag*., **27** 1231 (1991).

[8] J.O. Willis, R.D. Ray II, T.G. Holesinger, R. Zhou, K.V. Salazar, J.Y. Coulter, J.J. Gingert, D.S. Phillips and D.E. Peterson, "Bi-2212 and Bi-2223 Wire Development," *Proceedings of the 7th US/Japan Workshop on High Temperature Superconductors*, Tsukuba, Japan, 22-24 October 1995.

[9] J. Joo, J.P. Singh, T. Warzynski, A. Grow and R.B. Poeppel, "Role of Silver Addition on Mechanical and Superconducting Properties of High-T_c Superconductors," *Applied supercond.,* **2** [No.6] 401-410 (1994).

[10] T.E. Jones,W.C. McGinnis, E.W. Jacobs, R.D. Boss, P.M. Thibado, J.S. Briggs and W.E. Glad,"Effect of silver Additions on sintered $Bi_2Sr_2CaCu_2O_8$," *Physica C,* **201** 279-288 (1992).

[11] J.S. Luo, N. Merchant, V.A. Maroni, G.N. Riley, Jr. and W.L. Carter, "Influence of silver Cladding on the Formation and Alignment of the $(Bi_{2-x}Pb_x)Sr_2Ca_2Cu_3O_{10+\delta}$ Phase," *Appl. Phys. Lett.,* **63** (5) 690-692 (1993).

[12] S.X. Dou, H.K. Liu, Y.C. Guo, R. Bhasale, Q.Y. Hu, E. Babic and I.Kusevic, "Effect of Silver on the Processing and Properties of Bi2223/Ag Tapes," *Appl. Supercond.,* **2** [No 3/4] 191-199 (1994).

[13] K. Knizek, E. Pollert, D. Sedmidubsky, J. Hejtmanek ad J. Pracharova, "Single-Phase Region of the 2212-Bi-Sr-Ca-Cu-O Superconductor," *Physica C,* **216** 211-218 (1993).

[14] T.G. Holesinger, D.J. Miller and L.S. Chumbley, "Solid Solution Region of

the $Bi_2Sr_2CaCu_2O_y$ superconductor," *Physica C,* **217** 85-96 (1993).

[15] P. Majewski, "The High-T_c Superconducting Compounds of the System Bi-Sr-Ca-Cu-O," ,*Adv. Mater.* **4** 508 (1992); Adv. Mater. **6** 460 (1994).

[16] B. Hong, J. Hahn and T.O. Mason, "Phase Compositions and Compatibilities in the Bi-Sr-Ca-Cu Quarternary oxide system at 800°C in air," *J. Am. Ceram. Soc.,* **73** 1965 (1990).

[17] R. Muller, T. Schweizer, P. Bohac, R.O. Suzuki and L.J. Gauckler, "Compositional Range of the $Bi_2Sr_2CaCu_2O_x$ HT_c-Superconductor and its Surrounding Phases," *Physica C,* **203** (1992) 299-314.

[18] D.C. Sinclair, "Stoichiometry and Kinetics of Formation of $Bi_2Sr_2CaCu_2O_\delta$ Solid Solutions", *J. Mater. Chem.,* **2** (5) 579-580 (1992).

[19] P. Majewski, B. Freilinger, B. Hettich, T. Popp and K. Schulze, "Phase Equilibria in the system Bi_2O_3-SrO-CaO-CuO at Temperatures of 750°C, 800°C and 850°c in air," *High Temperature Superconductors - Materials Aspects,* ed. by H.C. Freyhardt et al., DGM Informations-gesellschaft-Verlag, **1** 393 (1991).

[20] J.I. MacManus-Driscoll and J.C. Bravman, "Effects of silver and lead on the Phase Stability of $Bi_2Sr_2CaCu_2O_{8+\delta}$ and $Bi_2Sr_2Ca_2Cu_3O_{10+\delta}$ above and below the Solidus Temperature," *J. Am. Ceram. Soc.,* **77** [9] 2305-13 (1994).

[21] J.I. MacManus-Driscoll, P-C Wang, J.C. Bravman, and R.B. Beyers, "Phase Equilibria and Melt processing of $Bi_2Sr_2CaCu_2O_{8-x}$ Tapes at Reduced oxygen partial pressures," *Appl. Phys. Lett.,* **22** 2872 (1994).

[22] K. Schulze, p. Majewski, B. Hettich, and Gunter Petzow, "Phase Equilibria in the system Bi_2O_3-SrO-CaO-CuO with emphasis on the High-T_c superconducting compounds," *Z. Metallkde,* **81** 836-842 (1990).

[23] B. Heeb, S. Oesch, P. Bohac, and L.J. Gauckler, "Microstructure of Melt-processed $Bi_2Sr_2CaCu_2O_x$ and Reaction Mechanisms During Post Heating Treatment," *J. Mater. Res.,* **7** [No. 11] 2948-2955 (1992).

[24] W. Wong-Ng, L.P. Cook and F. Jiang, "The Primary Crystallization Phase Field of "2212" Phase and the Effect of Ag Addition," *J. Amer. Ceram. Soc.,* to be published (1997).

[25] K.F.J. Heinrich, *Electron Beam X-ray Microanalysis,* Van Nostrand Reinhold Co., New York, 578 pp (1981).

[26] C.E. Fiori, C.R. Swyt, R.L. Myklebust, "NIST/NIH Desktop Spectrum Analyzer Program and X-ray Database," NIST Standard Reference Database No. 36 (1991).

[27] E.M. Levin,, C.R. Robbins, H. F. McMurdie, *Phase Diagrams for Ceramists,* Amer. Ceram. Soc., Columbus, OH, 600 pp (1964).

Pb-DISTRIBUTION IN A FIVE-PHASE (Bi,Pb)-Sr-Ca-Cu-O ASSEMBLAGE

Lawrence P. Cook and Winnie Wong-Ng
Ceramics Division
NIST
Gaithersburg
MD 20899

ABSTRACT

The partitioning of lead between phases in the nominal five-phase assemblage: $(Pb,Bi)_3Sr_2Ca_2CuO_x$ + $(Ca,Sr)_2PbO_4$ + $(Bi,Pb)_2Sr_2CaCu_2O_x$ + (Ca,Sr)O + liquid at 830°C in an atmosphere of 7.5 volume % O_2/ 92.5 volume % Ar has been investigated. Samples were studied using a combination of high temperature quenching, simultaneous differential thermal analysis and thermogravimetric analysis, scanning electron microscopy, and energy dispersive x-ray analysis. For the $(Pb,Bi)_3Sr_2Ca_2CuO_x$ and $(Ca,Sr)_2PbO_4$ phases, the average Pb/(Pb + Bi) molar ratios were determined to be 0.73 and 0.85, respectively. In the $(Bi,Pb)_2Sr_2CaCu_2O_x$ and liquid phases, the average Pb/(Pb+Bi) molar ratios were 0.14 and 0.24, respectively. This indicates an apparent saturation of Pb in the $(Bi,Pb)_2Sr_2CaCu_2O_x$ superconductor at a relatively low level. Data on Sr/(Sr+Ca) molar ratios for coexisting phases are also presented, along with a thermodynamically-derived projection of the five-component data for the coexisting phases.

INTRODUCTION

Although oxide superconductors such as $(Bi,Pb)_2Sr_2CaCu_2O_x$ (commonly referred to as "2212") and $(Bi,Pb)_2Sr_2Ca_2Cu_3O_x$ (or "2223") were originally discovered in the Pb-free Bi-Sr-Ca-Cu-O ("BSCCO") system [1-8], addition of lead was found to enhance the properties of wires and tapes processed from these materials [9-14]. The distribution of lead in superconductor-containing assemblages is an important aspect of the phase equilibria of such systems. Recently , we have studied the subsolidus five-phase equilibrium volumes involving the presence of coexisting 2223 and 2212 phases in the Bi-Pb-Sr-Ca-Cu-O ("Pb-BSCCO") system [15]. Sixteen such five-phase volumes were identified. Together, they form a close-packed compositional environment surrounding the 2223-2212 pair. In the present study, we describe results on lead partitioning between phases in a representative Pb-BSCCO five-phase volume, produced by partial melting of one these sixteen five-phase subsolidus volumes. Also as part of this study, we have used a thermodynamically-derived method for presenting the experimental results in this complicated five-oxide system.

EXPERIMENTAL METHODS

Crystalline starting materials for the experiments were prepared, via conventional calcining methods, using reagent grade powders of PbO, Bi_2O_3, $SrCO_3$, $CaCO_3$, and CuO. Initially, the following phases were synthesized: $(Bi,Pb)_2Sr_2CaCu_2O_x$, $(Bi,Pb)_2Sr_2Ca_2Cu_3O_x$, $(Ca,Sr)_2PbO_4$ (or "1x20"), $(Bi,Pb)_{11}Sr_9Ca_xCu_5O_y$ (or "119x5"), and $Bi_2Sr_3CaO_x$ (or "2310"). Approximately equal volumes of these phases were then mixed and equlibrated in an atmosphere of flowing 7.5 volume % O_2 / 92.5 volume % Ar at 815 °C - 820 °C for several days, with intermediate grindings. X-ray powder diffraction analysis showed that all five phases persisted, without the presence of new phases. Based on this information, the 2223 + 2212 + 1x20 + 119x5 + 2310 volume was considered to constitute a stable equilibrium assemblage.

To determine the initial melting temperature of this assemblage, a sample was subjected to simultaneous differential thermal analysis (DTA) and thermogravimetric analysis (TGA). For this experiment, the sample was placed in a MgO DTA crucible, on a thermocouple assembly calibrated against the melting point of gold. Powdered platinum was used as the reference. Temperatures were measured with an estimated standard uncertainty (one standard deviation) of less than ± 5°C. During the measurement, an atmosphere of 7.5 volume % O_2 / 92.5 volume % Ar was continuously flowed around the sample. The temperature was ramped at 4°C/min. The onset of melting was indicated by the rising DTA peak, with a corresponding weight loss monitored by TGA. The temperature ramp was stopped 5 °C above this onset, and held for 1 h, to allow the assemblage to attain equilibrium. Following this, the sample was furnace-cooled. Visual examination confirmed that melting had taken place. In a follow-up experiment, a sample of the unmelted assemblage was placed in a MgO crucible in contact with a porous MgO wick, according to a procedure previously described [16]. The sample was heated in a quench furnace 5°C above the DTA onset of melting in an atmosphere of flowing 7.5 volume % O_2 / 92.5 volume % Ar and quenched in liquid nitrogen-cooled helium gas.

Samples from both experiments were potted in epoxy, sectioned, polished and carbon-coated prior to microstructural and microchemical characterization using a scanning electron microscope (SEM) coupled with energy dispersive x-ray spectrometry (EDS). In these experiments, we used conventional methods of EDS data collection and reduction [17] via the software of Fiori et al [18]. Microanalytical standards were $PbZr_{0..41}Ti_{0.59}O_3$ for Pb and $Bi_2Sr_{1.5}Ca_{1.5}Cu_2O_x$ for the other elements. All analyses were completed at 20.3 kV. Uncertainties are estimated as follows: for mole fractions > 10 %, < ± 5 % relative expanded uncertainty; for mole fractions < 10 %, < ± 10 % relative expanded uncertainty.

RESULTS AND DISCUSSION

A backscattered electron micrograph of the DTA experiment, in which the 2223 + 2212 + 1x20 + 119x5 + 2310 assemblage was heated to 830°C, is shown in Figure 1a. A complex microstructure is apparent, and phase identification was made on the basis of EDS data. Most readily visible are the large four-sided crystals of the $(Pb,Bi)_3Sr_2Ca_2CuO_x$ phase (Bi-doped "451", commonly referred to

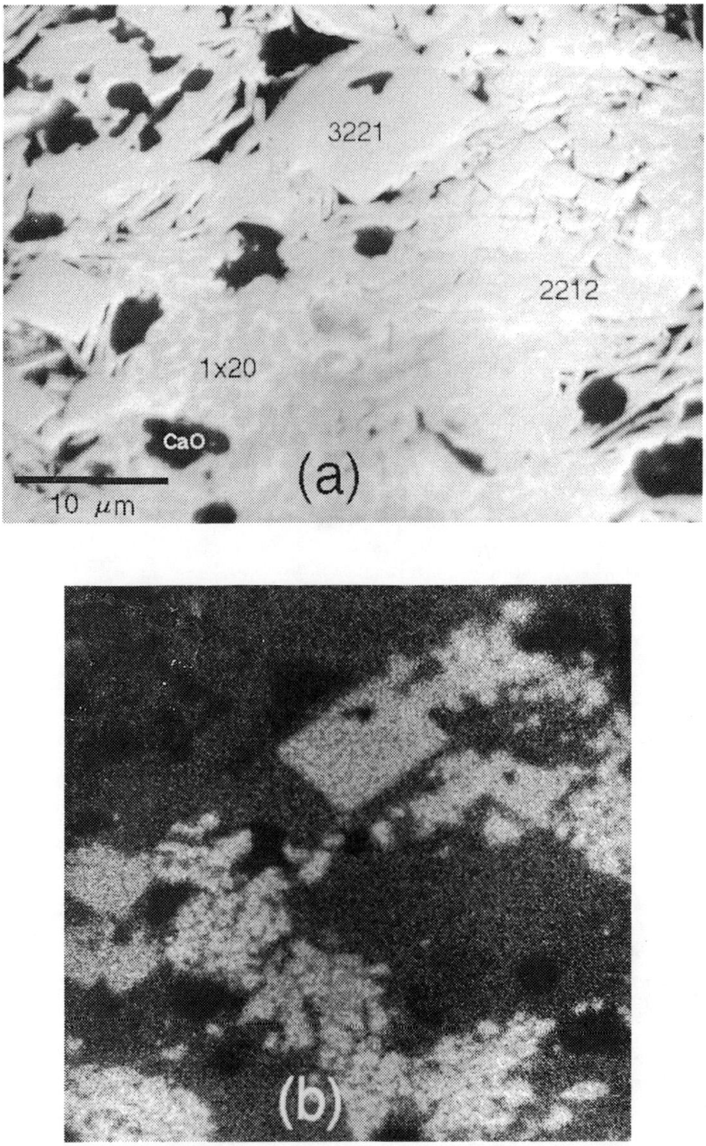

Figure 1. Sample of the initial assemblage: 2223 + 2212 + 1x20 + 119x5 + 2310 heated for 1 h at 830 C in 7.5 volume % O_2 / 92.5 volume % Ar.
a) Backscattered electron micrograph; b) x-ray map of Pb L:

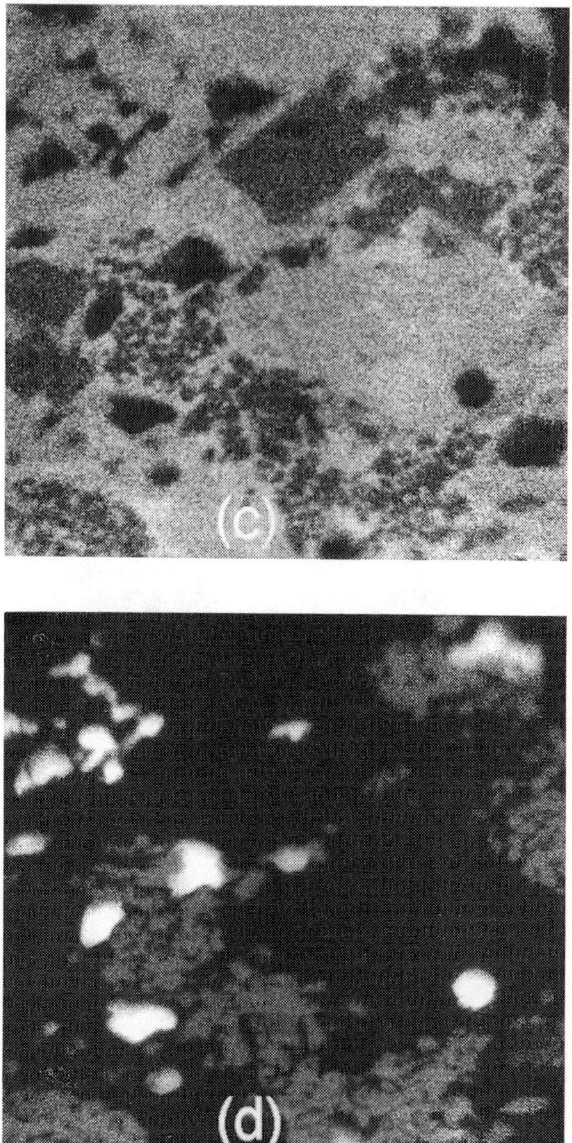

Figure 1, continued. c) x-ray map of Bi L; d) x-ray map of Ca K.

as "3221") at the upper center and middle left of the micrograph. X-ray maps showing the distribution of Pb, Bi and Ca are presented in Figs 1b, 1c and 1d, respectively. By comparison of these maps, three other principal phase regions, in addition to 3221 can be discerned. The clusters of relatively small grains, with strong Pb and moderate Ca intensities are 1x20. The areas having high Ca intensity are calcium oxide, which appears as dark, characteristically rounded areas in Fig. 1a. The areas of relatively low Pb and Ca intensity, with moderate Bi intensity, are 2212. Interstices between most of the 2212 crystals are filled with melt; however unfilled voids between these characteristically lath-like crystals can be seen at the lower right of Fig. 1a. Although recrystallized liquid is present and serves to hold the sample together, it is present only in thin areas between grains, and is not readily visible in the micrograph and x-ray maps of Fig. 1. As noted above, the presence of liquid was confirmed in a separate quenching experiment involving the use of a porous MgO wick. This infiltrated wick provided a sample of the melt for EDS analysis.

Initially the sample investigated was comprised of the five solid phases 2223, 2212, 1x20, 119x5, and 2310, yet melting of the sample created a new assemblage corresponding to 3221, 2212, 1x20, CaO and liquid. It is obvious that progressive melting in this multicomponent system is very complicated. For the present paper, we are interested in the distribution of lead between representative coexisting phases, and the phase assemblage of the sample selected for study is well suited for this purpose.

Several grains of each of the phases in Fig. 1 were analyzed by EDS. Averaged results of EDS analyses are given in Table Ia, along with calculated average stoichiometries for 3221, 1x20 and 2212 in Table Ib. For major components in Table Ia (mole fraction > 10 %), standard deviations are mostly in the 5 - 10 % (relative) range. This is somewhat greater than would be expected from the analytical uncertainty alone (refer to discussion above), and may be due to compositional variations among the grains analyzed. All the solid phases appear to contain measurable Bi, Pb, Sr, Ca, and Cu, suggesting the formation of complicated solid solutions, as there was no indication of inclusions in the grains selected for analysis. The average stoichiometry for the 3221 phase, as calculated in Table Ib on the basis of one Cu per formula, is substantially different than the nominal "3221". It is known that the 3221 phase encompasses a solid solution of considerable extent [19], although the limits have not been well defined. The average 1x20 stoichiometry of Table Ib indicates the Ca-rich end member, with only minor substitution of Bi for Pb. The average stoichiometry of the 2212 phase, as calculated in Table Ib, is reasonably close to the nominal, if entry of Ca and Pb into the Sr site is assumed. The CaO phase average of Table Ia contains 97 mol% Ca end member, with the balance being primarily Sr.

The complete set of EDS analyses, together with the averages of Table Ia, are plotted in Figure 2, which gives the Pb/(Pb + Bi) molar ratios as a function of total mole fraction (% $BiO_{1.5}$ + % PbO). For the 3221 and 1x20 phases, the average Pb/(Pb + Bi) molar ratios are 0.73 and 0.85 respectively. The average Pb/(Pb + Bi) molar ratios for the 2212 and liquid phases are 0.14 and 0.24, respectively. This may indicate a maximum Pb content of the 2212 superconductor under the conditions of the experiment, since the presence of 1x20 would imply saturation

Table Ia. Average phase compositions, from EDS analyses. Values indicated are mole fraction, %.[1]

Phase	3221	1x20	2212	CaO	liquid
# Analyses	10	10	11	5	1
$BiO_{1.5}$	8.0±1.1	4.1±1.2	25.1±1.2	0.6±0.2	23.6
PbO_x	21.7±0.9	23.9±1.4	4.3±0.6	0.3±0.2	7.6
SrO	33.3±1.6	7.5±1.5	25.3±1.9	1.3±0.4	26.3
CaO	25.8±1.9	61.6±2.9	16.8±1.4	97.1±0.9	25.4
CuO	11.2±0.8	2.9±1.4	28.5±1.5	0.8±0.3	17.1

[1] Stated uncertainties are one standard deviation, as derived from statistical analysis of multiple data points. Refer to text for discussion of overall relative expanded uncertainties.

Table Ib. Phase stoichiometries calculated according to structural formulas.

Phase	Average stoichiometry
3221	$(Bi_{0.71}Pb_{1.94})Sr_{2.97}Ca_{2.30}Cu_{1.0}O_x$
1x20	$(Ca_{1.85}Sr_{0.15})(Pb_{0.72}Bi_{0.12}Sr_{0.07}Cu_{0.09})O_4$
2212	$(Bi_{1.76}Pb_{0.30})Sr_{1.78}Ca_{1.18}Cu_{2.0}O_x$

with respect to the PbO component. As can be seen in Fig. 2, the scatter in analyses of 3221 and 1x20 phases extends in both the molar Pb/(Pb + Bi) and mole fraction (% $BiO_{1.5}$ + % PbO) directions. However, the 2212 analyses indicate relatively little variation in molar Pb/(Pb + Bi), with a substantial variation in mole fraction (% $BiO_{1.5}$ + % PbO). To fully treat such compositional effects, it is necessary to consider simultaneous variations in other elements, as well, since complex solid solutions are involved.

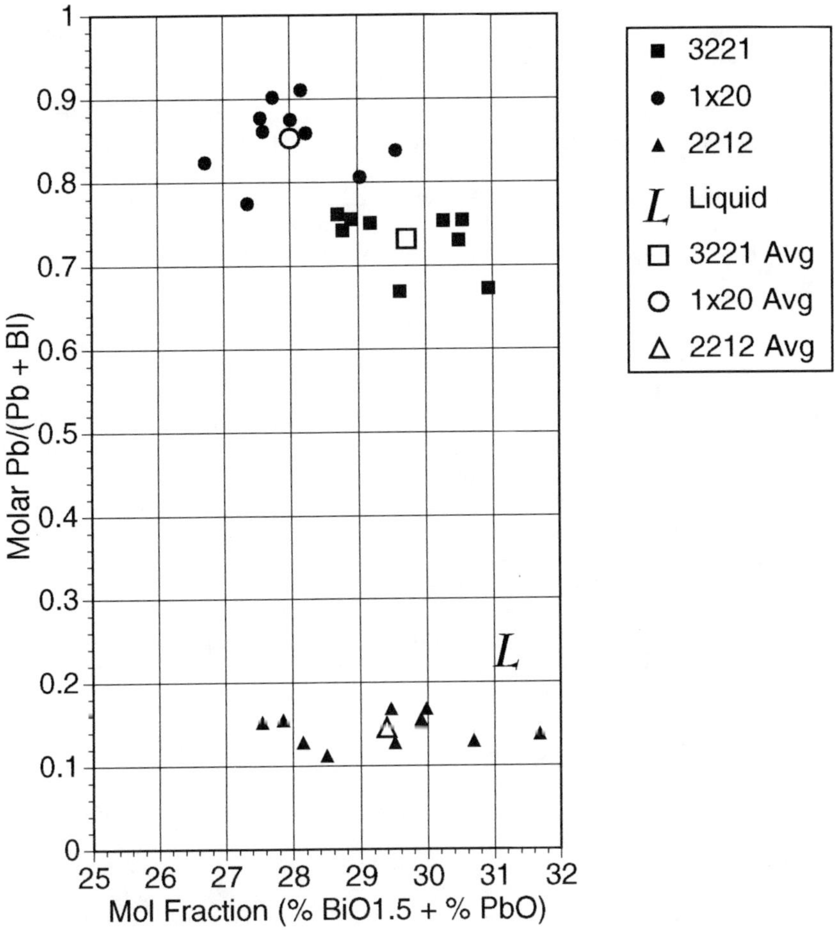

Figure 2. Plot of lead and bismuth data for five-phase assemblage of Fig. 1.

The molar ratio Sr/(Sr + Ca) is another useful parameter to consider, since it is well known that Sr and Ca substitute extensively for one another in many oxide solid solutions. This parameter is plotted in Figure 3, as a function of total mole fraction (% SrO + % CaO). In this diagram, it is seen that the molar Sr/(Sr + Ca) ranges of the 3221 and 2212 phases are similar, with averages (not shown) of 0.56 and 0.60, respectively. These averages are similar to that for the liquid phase, which is 0.51. The range for the 1x20 phase lies at a considerably lower value, with an average of 0.11. The CaO phase average, at molar Sr/(Sr + Ca) = 0.01, indicates the strongest partitioning of Ca. Comparison of the data in Figs. 2 and 3 indicates that the assumption of ideal substitution of Pb for Bi and Sr for Ca in the major Pb-containing phases is not a valid approximation. If it were, then there should be a continuous range of solid solutions. It appears that the solubility of Pb in 2212 is limited. Similarly the solubility of Bi in 3221 and 1x20 may be limited. The substitution of Ca for Sr and vica versa in the 3221 and 2212 structures is undoubtedly limited by the strong site preferences of these elements. For (Ca,Sr)O solid solutions, a relatively large miscibility gap has been reported at temperatures below 850 °C [20, 21].

A more complete graphical portrayal is essential for further treatment of the five-component phase composition data. Projective methods must be used to reduce the number of compositional variables for this purpose, as outlined in the following discussion.

GRAPHICAL REPRESENTATION OF DATA

The utility of phase equilibrium information is associated with its ready visualization in graphical form. However, the presentation of five-component Pb-BSCCO equilibria in graphical form is a major challenge. Several methods have been employed by various authors to reduce the dimensionality of their multicomponent phase equilibrium data. Among these are: 1) elimination of one or more components and renormalization in terms of the remainder (in effect, a "projection through" the components eliminated), 2) grouping of certain components and treating them collectively as one component, 3) sections through multicomponent space. The first method suffers from inaccuracies of representation associated with the fact that different compositions may plot at the same point. This can lead to serious ambiguities in the interpretation of data. The second method requires that the grouped elements substitute ideally for each other, leading to a reduction in the number of coexisting phases (these requirements are not met in the present case). The third method, while useful if the chosen compositional section passes through single-phase regions, results in a projection which can be very difficult to interpret in other regions of the phase diagram.

To avoid these difficulties, it is possible to devise a projection based on one or more of the solids of known composition present in the phase equilibrium assemblage. This method, which may be termed a "thermodynamically-derived" projection, differs from the others in that the point of reference is a phase in equilibrium with all others in the system. Therefore, there are no ambiguities in representation, as the straight-line nature of the tie-lines eliminates all possibilities for compositional overlap. This is a direct consequence of the thermodynamic

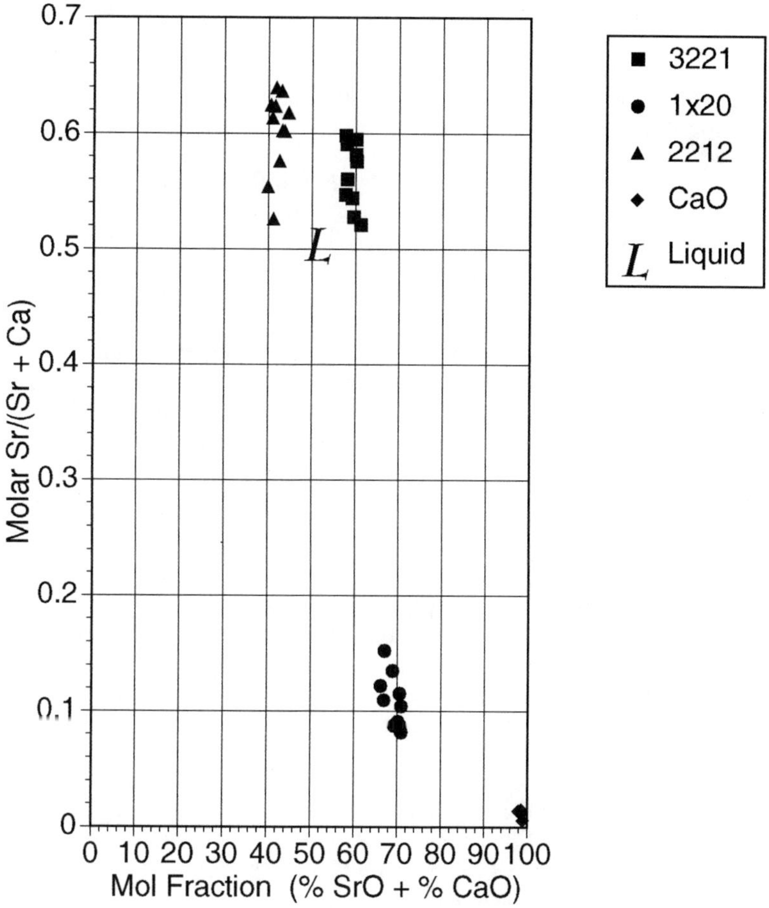

Figure 3. Plot of strontium and calcium data for five-phase assemblage of Fig. 1.

requirements for minimization of the Gibbs energy of the system. The phase diagram thus derived must be stipulated as containing the projective phase in equilibrium with all assemblages.

The thermodynamic method has been applied to the complete set of EDS data. Since the assemblage in the sample contains four solids, any of these could provide potential projection points. In principle, the solid used as the projection point should have a fixed composition. However, in the present situation, the inaccuracies associated with solid solution variation at the projection point are judged to be small relative to the advantage of being able to visualize the compositional interrelationships of the phases. We have determined that for the present set of data, a transformation from the original components $BiO_{1.5}$, PbO, SrO, CaO and CuO into a new component system comprised of 2212, PbO', 2310, CaO' and CuO' gives satisfactory results. The phase composition data may then be plotted unambiguously by using either 2212 or CaO as the projection point, since both are phases in equilibrium with the assemblage, and then plotting in terms of the other four components.

Figure 4 shows a projection through the 2212 phase, plotted in terms of PbO', 2310', CaO', and CuO'. The PbO' end member plots at the origin, and is not indicated on the axes. A more detailed description of the mechanism of the projection is given in Ref. [15]. Mathematically, the projection corresponds to a linear transformation between the original set of components, $BiO_{1.5}$, PbO, SrO, CaO, and CuO, and the new set of components. The new set of components were chosen as 2212 (e.g. $Bi_{2.0}Pb_{0.2}Sr_2CaCu_2O_x$), PbO', 2310 (i.e. $Bi_2Sr_3CaO_7$), CaO', and CuO'. It was found that this choice of components gives a projection with relatively low distortion. The new set of components is related to the original set according to the following, where the quantity "n" for each component refers to number of moles:

$$n_{2212} = 1.5(n_{BiO1.5}) - 1.0(n_{SrO})$$

$$n_{PbO'} = -0.3(n_{BiO1.5}) + 1.0(n_{PbO}) + 0.2(n_{SrO})$$

$$n_{2310} = -1.0(n_{BiO1.5}) + 1.0(n_{SrO})$$

$$n_{CaO'} = -0.5(n_{BiO1.5}) + 1.0(n_{CaO})$$

$$n_{CuO'} = -0.3(n_{BiO1.5}) + 2.0(n_{SrO}) + 1.0(n_{CuO}) \qquad (1)$$

The number of components in the new set is still five, but they have been defined so as to replace $BiO_{1.5}$ by 2212, thus enabling the projection through 2212. Plotting is achieved simply by normalizing compositions recalculated from equations [2] such that the mole fractions

$$PbO' + 2310 + CaO' + CuO' = 1.00 \qquad (2)$$

The composition of the 2212 used in the projection was that of Table Ib. Since

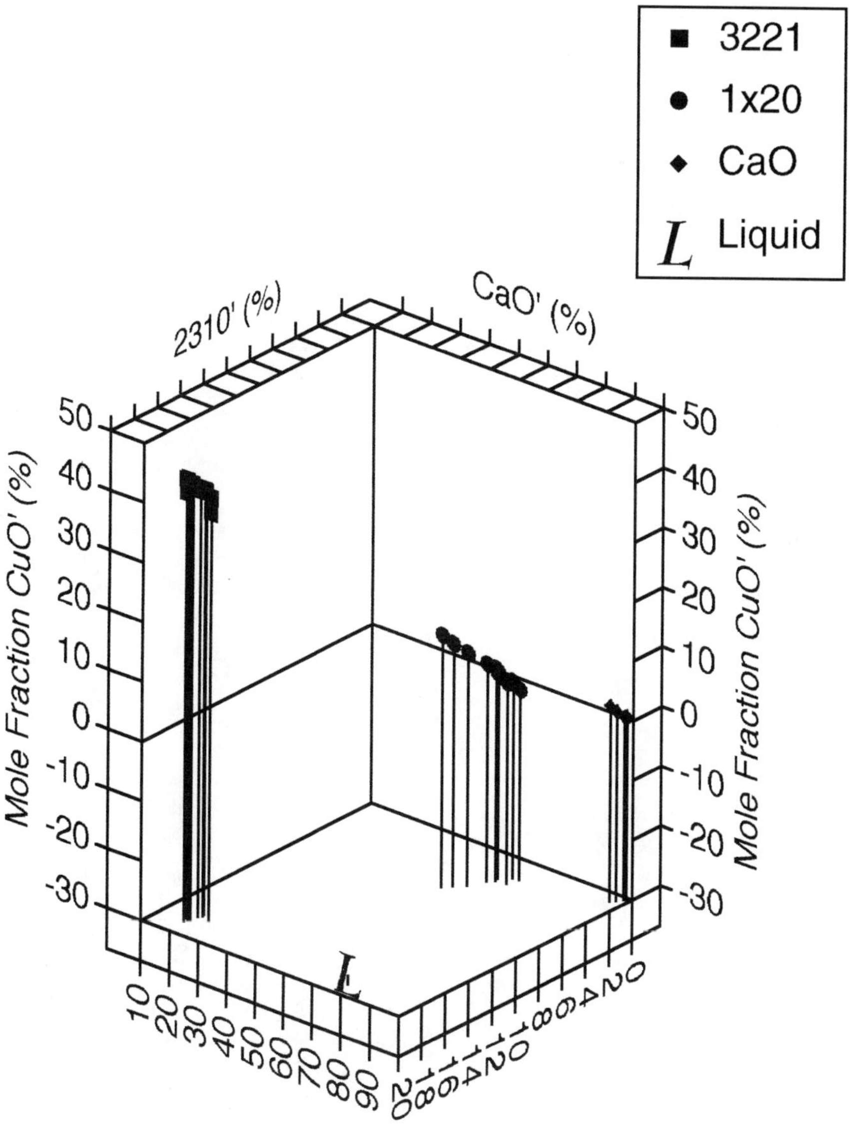

Figure 4. Projection of 3221, 1x20, CaO and liquid data through 2212.

the 2212 was used for the basis of the projection, it does not appear in the diagram of Fig. 4. The compositions of the other four phases are plotted, however. The phase regions of Fig. 4 show relatively regular trends. In particular, the 3221 and 1x20 regions, by comparison with Fig. 2, plot in nearly linear fashion. The compositional variations in 3221 and 1x20 appear to be more complex than involving simply Pb for Bi or Sr for Ca alone, and therefore they are not clearly displayed in Figs. 2 and 3, but rather require the more complete representation of Fig. 4 to become evident. The exact nature of the solid solution variations of Fig. 4 requires further study; nonetheless, the 3221 region is approximated by a line extending from $Bi_{9.8}Pb_{19.8}Sr_{31.8}Ca_{29.3}Cu_{9.3}O_x$ to $Bi_{7.1}Pb_{21.8}Sr_{35.7}Ca_{24.3}Cu_{11.1}O_x$, while the 1x20 region extends from $Bi_{3.8}Pb_{23.8}Sr_{5.8}Ca_{65.0}Cu_{1.6}O_x$ to $Bi_{3.5}Pb_{24.5}Sr_{9.3}Ca_{59.6}Cu_{3.1}O_x$. Coupled substitutions appear to be involved in both solid solutions. Use of the thermodynamic projection was instrumental in bringing to light the regularity of these variations.

Because 2212 was used as a projection point in Fig. 4, the 2212 phase region cannot be indicated in this diagram. However, since CaO is one of the phases present in the assemblage and is also one of the transformed components, it can be used as an alternative projection point, using the same transformation represented by equations (1). Plotting is achieved by normalizing the remaining components, according to:

$$2212 + PbO' + 2310 + CuO' = 1.00 \qquad (3)$$

When this is done, as in Figure 5, then the 2212 phase region can be portrayed. In Fig. 5, the 2212 phase region falls around an axis extending from the top rear vertex of the diagram down toward the front. It apparently forms a small elongated volume, rather than a line. This is consistent with the increased possibility for solid solution in the 2212 phase relative to the 1x20 phase. The 2212 solid solution region of Fig. 5 appears to extend (in terms of the 2212 structural formula) from approximately $(Bi_{1.62}Pb_{0.30})Sr_{1.88}Ca_{1.08}Cu_{2.0}O_x$ to $(Bi_{1.76}Pb_{0.26})Sr_{1.48}Ca_{1.34}Cu_{2.0}O_x$.

CONCLUSIONS

The addition of lead to the BSCCO system results in formation of the 3221 and 1x20 phases, which contain the majority of the lead, with relatively minor bismuth. The 2212 superconductor does not appear to take up a large amount of lead. At 830 °C, in an atmosphere of 7.5 volume % O_2/92.5 volume % Ar, and in the presence of 1x20, which implies near-saturation with respect to PbO, the 2212 was observed to contain lead corresponding to molar Pb/(Pb+Bi) = 0.14. The liquid phase contained slightly more lead, corresponding to molar Pb/(Pb + Bi) = 0.24. By projecting through 2212, a graphical representation of the five-component Pb-BSCCO system can be obtained. This projection is useful for assemblages containing 2212 in order to portray the compositional relationships and tie lines of the coexisting phases. For superconductor processing, the implication is that addition of a mole fraction of only 5 - 10 % PbO, depending on the bulk composition and the secondary phases desired, may be sufficient to saturate the superconductor phase.

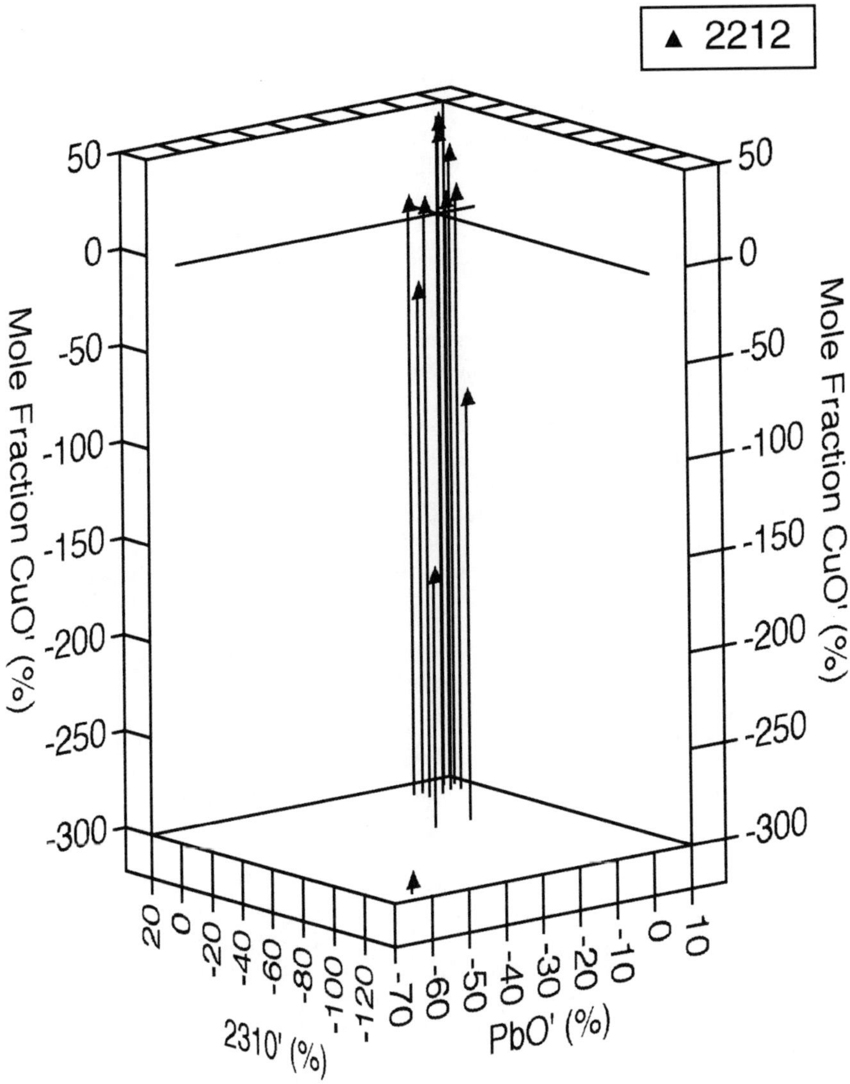

Figure 5. Projection of 2212 data through CaO .

ACKNOWLEDGEMENTS

Partial support for this study was provided by the U.S. Department of Energy. The sample was prepared by William Greenwood of the University of Maryland.

REFERENCES

[1] H. Maeda, Y. Maeda, M. Tanaka, M. Fukutomi and T. Asano, "A new High-T_c Oxide Superconductor without a Rare Earth Element," *Japan J. Appl. Phys.* **27** [2] L209-L210 (1988).

[2] L.M. Rubin, T.P. Orlado, and J.B. Vander Sandre, G. Gorman, R. Savoy, R. Swope, and R. Beyers, "Phase Stability Limits of $Bi_2Sr_2CaCu_2O_{8+x}$and $Bi_2Sr_2Ca_2Cu_3O_{10+x}$" *Appl. Phys Lett.* **61**16] 1977 (1992).

[3] P.V.P.S.S. Sastry and A.R. West, "Synthesis, Stoichiometry and T_c of the Pb-Free $Bi_2Sr_2Ca_2Cu_3O_{10}$ Superconductor'" *Physica C*, **232** 63-68 (1994).

[4] K. Knizek, E. Pollert, D. Sedmidubsky, J. Hejtmanek ad J. Pracharova, "Single-Phase Region of the 2212-Bi-Sr-Ca-Cu-O Superconductor," *Physica C*, **216** 211-218 (1993).

[5] T.G. Holesinger, D.J. Miller and L.S. Chumbley, "Solid Solution Region of the $Bi_2Sr_2CaCu_2O_y$ superconductor," *Physica C*, **217** 85-96 (1993).

[6] P. Majewski, "The High-T_c Superconducting Compounds of the System Bi-Sr-Ca-Cu-O," *Adv. Mater.*, **4** 508 (1992); *Adv. Mater.*, **6** 460 (1994).

[7] R. Muller, T. Schweizer, P. Bohac, R.O. Suzuki and L.J. Gauckler, "Compositional Range of the $Bi_2Sr_2CaCu_2O_x$ HT_c-Superconductor and its Surrounding Phases," *Physica C*, **203** (1992) 299-314.

[8] D.C. Sinclair, "Stoichiometry and Kinetics of Formation of $Bi_2Sr_2CaCu_2O_{8+x}$ Solid Solutions", *J. Mater. Chem.*, **2** (5) 579-580 (1992).

[9] S. Kesche, P. Majewski, and F. Aldinger, "Phase Relations and Homogeneiiy Region of the High Temperature superconducting phase $(Bi,Pb)_2Sr_2Ca_2Ca_3O_{10+d}$," *J. Elect. Mater.*, **24** [12] 1829 (1995).

[10] S. Kesche, P. Majewski, and F. Aldinger, "Phase equilibria in the System Bi_2O_3-PbO-SrO-CaO-SrO with Special Regard to the $(Bi,Pb)_{2+x}Sr_2Ca_2Cu_3O_{10+d}$ phase," *Z. Metallkd.*, to be published, (1997).

[11] K.H. Sandhage, G.N. Riley, Jr., and W. Carter, "Critical Issues in the OPIT Processing of High-J_c BSCCO Superconductors," *J. Metals*, **43** 21 (1991).

[12] R.W. McCallum, K.W. Dennis, L. Margulies and M.J. Kramer, "Processing and Properties of Long lengths of Superconductors", *Proceedings of the 1993 Falls TMS meetings-Pittsburgh, PA*, October, 17-21 (1993).

[13] J. I. MacManus-Driscoll and J. C. Bravman, "Effects of silver and lead on the Phase Stability of $Bi_2Sr_2CaCu_2O_{8+x}$and $Bi_2Sr_2Ca_2Cu_3O_{10+x}$above and below the Solidus Temperature," *J. Am. Ceram. Soc.*, **77** [9] 2305-13 (1994).

[14] U. Balachandran and A. N. Iyer, P. Haldar, J.G. Hoehn, Jr. and L. R. Motowidlo, "Progress in Development of Tapes and Magnets Made From Bi-2223 Superconductors," *Proceedings of The Fourth International conference ad Exhibition: World Congress on Superconductivity*, Vol. II, edited by K. Kristen and C. Burnham, Orlando, FL. June 27-July 1, 1994, pp. 639-649.

[15] W. Wong-Ng, L. P. Cook, F. Jiang*, W. Greenwood*, U. Balachrandran,

and M. Lanagan, "Subsolidus Phase Equilibria of the High Tc Pb-2223 Superconductor in the (Bi,Pb)-Sr-Ca-Cu-O System under 7.5% O2, " to be published, *J. Mater. Res.*, (1997).

[16] W. Wong-Ng and L. P. Cook, "Oxidation/Reduction Melting Equilibria in the System BaO-1/2Y_2O_3-CuO_x. I. Methods", p. 73-84 in K. M. Nair, U. Balachandran, Y-M. Chiang, and A. Bhalla, eds., Superconductivity and Ceramic Superconductors II, *Ceramic Trans.* **18**, Amer. Ceram. Soc., Westerville, OH (1991).

[17] K.F.J. Heinrich, *Electron Beam X-ray Microanalysis*, Van Nostrand Reinhold Co., New York, 578 (1981).

[18] C.E. Fiori, C.R. Swyt, R.L. Myklebust, "NIST/NIH Desktop Spectrum Analyzer Program and X-ray Database," NIST Standard Reference Database No. 36 (1991).

[19] H. Kitaguchi, T. Jun, K. Oda and Y. Miura, "Equilibrium Phase Diagrams for the Systems PbO-SrO-CuO and PbO-CaO-SrO," *J. Mater. Res.*, 5 [7] 1397-1402 (1990).

[20] R.S. Roth, "User aspects of phase equlibria in high T_c ceramic superconductors"; in Proc. User Aspects of Phase Equilibria, Joint Research Centre, Petten, Netherlands, Institute of Metals, London, England, June 25-27 (1990).

[21] W. Wong-Ng, L.P. Cook, F. Jeng and W. Greenwood, "Phase Relationships of the Pb-Sr-Ca-O system in air," *Physica C*, 272 [No 1 & 2] 87-93 (1996).

produced superconductors with excellent properties, most notably high critical current density [1-3]. Consequently, efforts have concentrated on producing superconducting tapes, wires, and coils of these materials, with emphasis on the Pb-free 2212 and Pb-doped 2223 ([Bi,Pb]:Sr:Ca:Cu) phases [4,5].

Phase equilibrium data are essential for guiding the processing of high-T_c phases. In particular, melting data are needed for PIT processing, and melt growth of single crystals. While melt data are relatively plentiful for the Pb-free 2212 phase [6-8], only limited data are available for the Pb-doped 2223 phase [9-13]. The goal of this paper is to report the progress of our research on the melting investigation of the Pb-2223 phase, including melting temperatures of selected five-phase volumes as obtained from differential thermal analysis (DTA) measurements, and the corresponding melt compositions as obtained from scanning electron microscopy and energy dispersive X-ray spectrometry (SEM/EDS).

The method of locating the primary phase field has been illustrated and discussed in detail previously with reference to the Pb-free 2212 phase [6]. Analogously, the first step is determination of all phases that are in equilibrium with the Pb-2223 phase. Next, the equilibrium five-phase combinations that include these phases with 2223 are determined. Becuase there will be a potentially large number of such equilibrium volumes, we have initially narrowed the search by sequentially considering volumes that include 2223 and a designated second phase as a pair. So far, we have concentrated our effort on the study of the five-phase volumes that include three pairs of phases: 2223 + 2212, 2223 + CuO, and 2223 + $(Ca,Sr)_2CuO_3$. During the 2223-phase formation process, the 2212, CuO, and $(Ca,Sr)_2PbO_4$ phases are present as precursors [14,15]. Also, the processing of the 2223 phase frequently leads to the presence of these three phases as impurity phases. Therefore, a precise knowledge of the phase relationships of the 2223 phase with 2212, CuO, and $(Ca,Sr)_2PbO_4$ is of significant value for 2223 processing.

II. EXPERIMENTAL METHODS

The samples for this investigation were prepared by the solid state calcining technique. Stoichiometric mixtures of PbO, Bi_2O_3, $SrCO_3$, $CaCO_3$ and CuO were pressed and heat treated at temperatures within the stable range of the desired phases, according to available phase diagram information [16]. Approximately 30 g of the Pb-2223 phase were prepared by using $Bi_{1.8}Pb_{0.4}Sr_2Ca_{2.2}Cu_3O_x$. The heat treatment process for this phase was as follows: 840°C (24 h), 850°C (24 h), 855°C (40 h), 860°C (40 h), and 855°C (120 h).

As reported elsewhere [17], 11 phases were found to be in equilibrium with the Pb-2223 phase. Also, Pb was found to substitute in the Bi-containing compounds that are in equilbrium with the Pb-2223 phase. Samples that represent five-phase combinations were prepared by mixing approximately equal volumes of these phases, pelletizing, and calcining at 810-820°C for **2** days in a 7.5% O_2 (92.5% Ar) atmosphere, followed by grinding, repelletizing, and further reheating at the same temperature for another 2 days. All samples were subjected to powder X-ray diffraction to confirm that the five- phase compatibilities persisted. Simplified abbreviations are used to designate the complicated formulas of these starting compounds, most of which are solid solutions, as shown in Table I.

Table I. Symbols and compositions of compounds used in this study

Symbol	Composition
2223	$(Bi_{1.8}Pb_{0.4})Sr_2Ca_{2.2}Cu_3O_x$
2212	$Bi_{1.9}Pb_{0.1}Sr_{1.5}Ca_{1.5}Cu_2O_x$
014x24	$Sr_7Ca_7Cu_{24}O_{41}$
0x21	$(Ca_{1.9}Sr_{0.1})CuO_3$
3221	$Bi_{0.5}Pb_3Sr_2Ca_2CuO_x$
2310	$Bi_{27.8}Pb_{6.18}Sr_{49.5}Ca_{16.5}O_x$
2201	$Bi_{1.64}Pb_{0.36}Sr_2CuO_x$
119x5	$Bi_{1.8}Pb_{0.4}Sr_{1.6}Ca_{0.2}CuO_x$
0x11	$(Ca_{0.86}Sr_{0.14})CuO_2$
0x11'	$(Ca_{0.5}Sr_{0.5})CuO_2$
1x20	$(Ca_{1.9}Sr_{0.1})PbO_4$
CaO	(Ca,Sr)O

The procedure for obtaining compositions of melts that are produced during initial melting of multiphase volumes has been documented elsewhere [6]. The various steps can be summarized as follows: (1) DTA/thermogravimetric analysis studies to obtain an indication of thermal events, (2) annealing in purified air and quenching of samples in liquid- nitrogen-cooled helium for further characterization, (3) powder x-ray characterization of solid residual phases that are present and of selected wick material to identify the crystallized melt (see Step 5), (4) SEM and X-ray mapping to study the microstructure of the quenched materials, (5) equilibration with porous MgO wicks to capture the liquid that is formed, and (6) quantitative SEM/EDS to determine the composition of the quenched-wick experiments. We accomplished the latter by established methods [18] in conjunction with the software DTSA [19], which incorporates several advanced features for spectral manipulation and quantification. The standards used for microanalysis were $Bi_2Sr_{1.5}Ca_{1.5}Cu_2O_x$ and (Pb, Zr)TiO_3. Analytical uncertainties (one standard deviation) are estimated at $<10\%$ relative. DTA temperatures were measured during the heating cycle at a scan rate of 4°C/min. Uncertainties in temperatures reported in this paper are estimated at $\pm$ 7°C.

III. RESULTS AND DISCUSSION

A. Subsolidus

The compounds in the Bi-Pb-Sr-Ca-Cu-O system include a complicated series of binary, ternary, quarternary, and five-component solid solutions. Therefore, a large number of two phase tie-line bundles, and stacks of three-phase tie planes exist. In the following discussion, the exact equilibrium compositions of the solid solutions that participate in specific equilibria are not specified. However, it should be noted that these compositions, although not yet experimentally determined, are always thermodynamically defined in this system for equilibrium assemblages that contain five phases.

From the X-ray diffraction results, 11 phases were found to be in equilibrium with the Pb-2223 phase at 810°C-820°C [17]. Phases in equilibrium with the Pb-2223 phase are 2212, CuO, CaO, 014x24, 0x21, 3221, 2310, 119x5, 0x11', 0x11, 1x20.

Table II lists the 16 five-phase volumes that contain the 2223 + 2212 pair at 810- 820°C in 7.5% O_2. These subsolidus volumes are consistent with each other in that each volume shares a given side with no more than one other. Using the same principle, we identified nine five-phase volumes that contained the 2223 + CuO pair, as indicated in Table III. In addition, we have determined four five-phase volumes that contained the 2223 + $(Ca,Sr)_2CuO_3$ pair, as given in Table IV. The number of volumes for the latter two pairs is much smaller than for the 2223 + 2212 pair. This is so because CuO and $(Ca,Sr)_2CuO_3$ are one- and three-component phases, respectively, whereas 2212 is a five-component phase (when it contains lead) that lies nearer the center of the system. This gives it access to tie lines with a much larger number of phases. The visual representation of the 16 subsolidus volumes that contain the 2223 + 2212 pair was discussed previously [17]. Although similar methods could be applied to volumes that contain the other pairs, this is currently under development and will be presented in a future paper. The stability of five-phase volumes based on other pairs, including 2223, is under investigation.

Table II. Five-phase equilibrium volumes that contained the 2212 + 2223 pair in 7.5% O_2 [17]. DTA temperatures indicate iniitial melting, quench (Qch) temperatures indicate temperature of melt wick sampling experiments.

Five-phase Equilibrium[a]	DTA T(°C)[b]	Qch T(°C)[c]	Melt Composition (Mole %)				
			$BiO_{1.5}$	PbO	SrO	CaO	CuO
2223-2212-1x20-119x5-2310	825	830	23.6	7.6	26.3	25.4	17.1
2223-2212-1x20-2310-CaO	830	835	23.0	7.3	27.1	23.5	19.1
2223-2212-0x21-1x20-CuO	837	840	18.5	14.4	16.3	14.2	36.6
2223-2212-1x20-0x11-3221	838	842	19.9	14.0	19.9	15.4	30.8
2223-2212-1x20-CuO-0x11	839	842	15.0	10.0	16.0	12.0	47.0
2223-2212-0x21-014x24-CuO	840	850	28.0	3.2	17.1	21.6	30.1
2223-2212-014x24-1x20-119x5	840	845	20.2	6.5	20.8	22.2	30.3
2223-2212-2310-0x11-3221	840	845	13.5	4.1	27.8	26.4	28.2
2223-2212-119x5-0x11-CuO	842	846	17.2	2.7	19.9	28.7	31.5
2223-2212-014x24-119x5-CuO	845	850	21.4	2.8	24.0	20.9	30.9
2223-2212-014x24-0x21-1x20	846	851	17.0	11.0	16.2	17.9	37.9
2223-2212-3221-CaO-1x20	848	852	19.3	10.2	23.7	18.4	28.4
2223-2212-0x11-2310-119x5	850	853	21.6	4.7	24.5	26.2	23.0
2223-2212-0x11-0x11'-CuO	850	855	21.8	1.2	21.7	18.3	37.0
2223-2212-0x11-0x11' 2310	853	858	25.3	2.3	22.7	17.3	32.4
2223-2212-3221-CaO-2310	865	870	20.4	7.6	20.9	25.8	25.3

[a] See Table I for symbols and compositions

[b] indicate initial melting

[c] indicate temperature of melt wick sampling experiments

B. Initial Melting

The DTA temperatures of initial melting for each of the five-phase volumes are listed in Tables II - IV, where they are arranged in ascending order. As was observed for the Pb-free 2212 [6] and 2201 phases [20], the Pb-2223 phase melts incongruently [9]. In 7.5% O_2, the initial melting temperatures for the five-phase volumes that have been determined ranged from 815°C to 865°C. This spread of 50°C is somewhat smaller than the ≈70°C spread observed in the four-phase volumes that contain the Pb-free 2212 phase [6].

To further characterize the initial melting of the 16 five-phase volumes that contained the 2212 +2223 pair, annealing and quenching experiments, in which a porous MgO wick was included, were conducted at temperatures a few degrees above the DTA temperature, as shown in Table II. The compositions of initial melts, based on EDS analyses of the MgO wicks, are indicated in Table II. By analogy with our determination of the primary phase volume of the Pb-free 2212 phase [6], the 16 compositions in Table II describe the extent of Pb-2223 crystallization volume for the portion that includes the 2223 + 2212 pair. In principle, when a composition is prepared within these limits, crystals of 2223 and 2212 together will be the first to crystallize from the melt.

The range of melt compositions expressed in mole % is $BiO_{1.5}$,15.0-28.0; PbO, 1.2-14.4; SrO, 16.0-27.1; CaO, 12.0-28.7; and CuO, 17.1-47.0. These are the limits of the primary phase field of the Pb-2223 phase in coexistence with 2212. This multicomponent 'hypervolume' cannot be depicted directly. One approximate way to present these data is via projections. Figure 1 shows the projected limits of the melt compositions, as plotted in a series of triangular diagrams with ½Bi_2O_3-CaO-CuO, ½Bi_2O_3-SrO-CuO, ½Bi_2O_3-PbO-CuO, ½Bi_2O_3-PbO-CaO, ½Bi_2O_3-PbO-SrO, and ½Bi_2O_3-SrO-CaO at the corners. In these projections, the compositions are normalized with respect to the three designated oxides. Because overlap is possible, the projections do not provide an unambiguous representation, but they are, nonetheless, useful to qualitatively describe the relative extent of the partial crystallization field. Although PbO is present in the liquid of every volume, the concentration of PbO is the least among the five oxide components, followed by SrO. The concentration of the CuO component was the highest concentration. Referring to Table II, the subsolidus volumes with relatively high PbO concentration give rise to higher PbO content in the melt. For example, in the five-phase volume 2223-2212-1x20-0x11-3221, which contains the PbO-rich phases $Bi_{0.5}Pb_3Sr_2Ca_2CuO_x$ and $(Ca_{1.9}Sr_{0.1})PbO_4$, the PbO concentration is relatively high (14.0 %). By contrast, for the volume 2223-2212-0x11-0x11'-CuO, where there is no PbO-containing compound other than a small amount of Pb substituted in the 2223, 2212 and 119x5 phases, the PbO content of the liquid is only 1.2 %.

Table III. Initial melting temperatures of nine five-phase volumes that involve the 2223 + CuO pair in 7.5% O_2.

Five-phase equilibrium	DTA T (°C)
2223-CuO-2212-0x21-1x20	837
2223-CuO-2212-1x20-0x11	839
2223-CuO-2212-0x21-014x24	840
2223-CuO-2212-119x5-0x11	842
2223-CuO-2212-014x24-119x5	845
2223-CuO-2212-0x11-0x11'	850
2223-CuO-1x20-0x21-014x24	828
2223-CuO-1x20-0x11'-014x24	815
2223-CuO-0x11'-1x20-0x11	838

Table IV. Initial melting temperatures of four five-phase volumes involving the 2223 + $(Ca,Sr)_2CuO_3$ pair under 7.5% O_2

Five-phase equilibrium	DTA T (°C)
2223-0x21-2212-1x20-CuO	837
2223-0x21-2212-014x24-CuO	840
2223-0x21-2212-014x24-1x20	846
2223-0x21-1x20-CuO-014x24	828

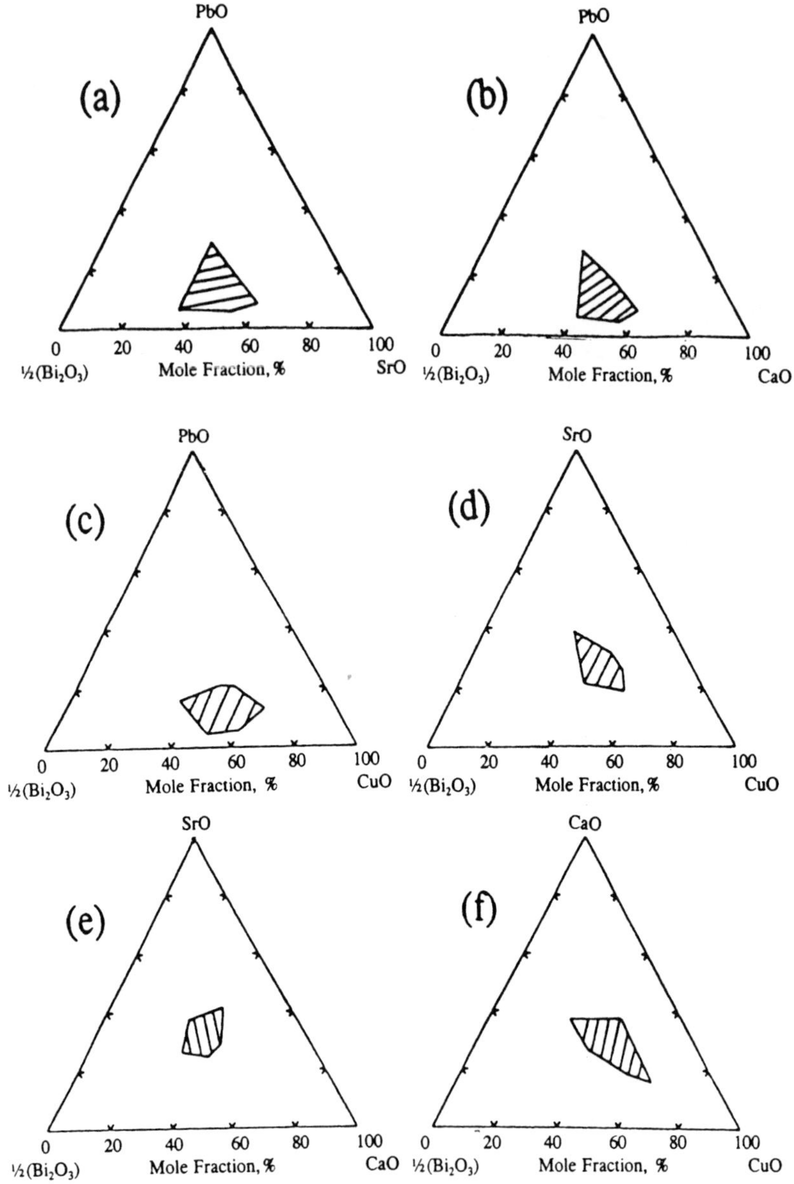

Figure 1. Projections that show approximate extent of the Pb-2223 + 2212 crystallization field with respect to various combinations of normalized

components.

IV. SUMMARY

The subsolidus phase equilibria and liquid melt compositions of a subset of the five-phase volumes that contain the Pb-2223 phase were investigated. Sixteen five-phase volumes that contained the 2223 + 2212 pair, nine volumes that contained the 2223 + CuO pair, and four volumes that contained the 2223 + 0x21 pair have been found. These volumes are consistent with each other. The liquid compositions of the initial melts of the five-phase volumes involving the 2223 + 2212 pair have been measured. It was found that, for these 16 volumes, the range of initial melt compositions encompasses $BiO_{1.5}$, 15.0-28.0; PbO, 1.2-14.4; SrO, 16.0-27.1; CaO, 12.0-28.7; and CuO, 17.1- 47.0. A PbO component is present in the initial melt of every volume. The subsolidus volume that contains PbO-rich phases gives rise to a melt with the highest PbO content.

V. REFERENCES

[1]U. Balachandran and A. N. Iyer, P. Haldar, J.G. Hoehn, Jr. and L. R. Motowidlo, "Progress in Development of Tapes and Magnets Made From Bi-2223 Superconductors," *Proceedings of The Fourth International conference ad Exhibition: World Congress on Superconductivity*, Vol. II, edited by K. Kristen and C. Burnham, Orlando, FL. June 27-July 1, 1994, pp. 639-649.

[2]K. H. Sandhage, G. N. Riley, Jr., and W. Carter, "Critical Issues in the OPIT Processing of High-J_c BSCCO Superconductors," *J. Metal*, **43** 21 (1991).

[3]K. Sato, T. Hikata, H. Mukai, M. Ueyama, N. Shibata, T. Kato, T. Masuda, M. Nagata, K. Iwata and T. Mitsui, *IEEE Trans Mag.*, **27** 1231 (1991).

[4]A. P. Malozemoff, Q. Li, and S. Fleshler, "Progress in BSCCO-2223 Tape Technology," submitted, 1997.

[5]J. O. Willis, R. D. Ray II, T. G. Holesinger, R. Zhou, K. V. Salazar, J. Y. Coulter, J. J. Gingert, D. S. Phillips, and D. E. Peterson, "Bi-2212 and Bi-2223 Wire Development," *Proceeding of the 7th US/Japan Workshop on High Temperature Superconductors*, Tsukuba, Japan, 22-24 October 1995.

[6]W. Wong-Ng, L. P. Cook, and F. Jiang, "The Primary Crystallization Phase Field of "2212" Phase and the Effect of Ag Addition," *J. Amer. Ceram. Soc.*, to be published (1997).

[7]P. Majewski, "The High-T_c Superconducting Compounds of the System Bi-Sr-Ca-Cu-O," *Adv. Mater.*, **4** 508 (1992).

[8]J. I. MacManus-Driscoll and J. C. Bravman, "Effects of silver and lead on

the Phase Stability of $Bi_2Sr_2CaCu_2O_{8+\delta}$ and $Bi_2Sr_2Ca_2Cu_3O_{10+\delta}$ above and below the Solidus Temperature," *J. Am. Ceram. Soc.,* **77** [9] 2305-13 (1994).

[9]S. Kesche, P. Majewski, and F. Aldinger, "Phase Relations and Homogeneiiy Region of the High Temperature superconducting phase $(Bi,Pb)_2Sr_2Ca_2Ca_3O_{10+d}$," *J. Elect. Mater.*, **24** [12] 1829 (1995).

[10]S. Kesche, P. Majewski, and F. Aldinger, "Phase equilibria in the System Bi_2O_3-PbO-SrO-CaO-SrO with Special Regard to the $(Bi,Pb)_{2+x}Sr_2Ca_2Cu_3O_{10+d}$ phase," *Z. Metallkd.*, to be published, (1997).

[11]Y. S. Sung and E. E. Hellstrom, "A Role of Ag in the Reaction Forming $(Bi,Pb)_2Sr_2Ca_2Cu_3O_x$ from $(Bi,Pb)_2Sr_2CaCu_2O_x$, $(Sr,Ca)_2CuO_3$ and CuO," *Physica C,* **255**, 266-274 (1995).

[12]K. Osamura and T. Maruyama, "Phase Relation in Ag/Bi2223 Tapes," *Proceeding of the Society of Non-traditional Technology*, organized by the New Superconducting Materials Forum, Tsukuba, 1 (1992).

[13]Y. S. Sung and E. E. Hellstrom, "A Reaction Couple Study of Reactions Amongst $Bi_2Sr_2CaCu_2O_x$, Ca_2PbO_4, and CuO to form $(Bi,Pb)_2Sr_2Ca_2Cu_3O_x$," *Physica C*, **253** 79-88 (1995).

[14]W. Wong-Ng and S. W. Freiman, "Superconducting Phase Formation in Bi(Pb)-Sr-Ca-Cu-O Glasses: A Review," *Superconducting Glass-Ceramics in BSCCO: Fabrication and Its Application*, to be published, 1997.

[15]W. Wong-Ng and S. W. Freiman, "High T_c Superconducting Bi-Sr-Ca-Cu-O Glass Ceramics: A Review," *Appl. Supercond.*, **2** (3/4) 163 (1994).

[16]J. D. Whitler and R. S. Roth, *Phase Diagrams for High T_c Superconductors*, The Amercian Ceramic Society, Westerville, OH (1991).

[17]W. Wong-Ng, L. P. Cook, F. Jiang, W. Greenwood, U. Balachrandran, and M. Lanagan, "*Subsolidus Phase Equilibria of the High T_c Pb-2223 Superconductor in the (Bi,Pb)-Sr-Ca-Cu-O System Under 7.5% O_2*," submitted to J. Mater. Res. (1997).

[18]K. F. J. Heinrich, *Electron Beam X-ray Microanalysis*, Van Nostrand Reinhold Co., New York, p.578 (1981).

[19]C. E. Fiori, C. R. Swyt, and R. L. Myklebust, *NIST/NIH Desktop Spectrum Analyzer Program and X-ray Database*, NIST Standard Reference Database No. 36 (1991).

[20]W. Wong-Ng, L. P. Cook, and F. Jiang, "Solid solution Region and Melting characterization of $Bi_{1.8+x}Sr_{2.2-x-y}Ca_yCu_{1\pm z}O_w$," to be submitted to *Physica C*, (1997).

Crystal Growth and Synthesis

XRD ANALYSIS OF YBCO SUPERCONDUCTING FILMS

E. J. Peterson, P. N. Arendt, J. Y. Coulter, P. C. Dowden, S. R. Foltyn, J. R. Groves , W. L. Hults, J. L .Smith, and J. O. Willis, Los Alamos National Laboratory, Los Alamos, NM 87545

ABSTRACT

X-ray diffraction (XRD) analysis was performed on samples of $YBa_2Cu_3O_{7-\delta}$ (YBCO) thin films that had been deposited by pulsed laser deposition at fast and slow rates on $SrTiO_3$ single crystal substrates. As deposited samples were given subsequent high temperature post anneal treatments. While oxygen stoichiometry most likely remained essentially constant at fully oxygenated values in these samples, c-axis lattice parameter and nonuniform strain values obtained from the XRD analysis were found to closely track critical current density values.

INTRODUCTION

Over the past decade, much effort has been focused on the development and commercialization of high temperature superconducting ceramics for electrical power and electronic device applications. The first generation of high T_c bulk conductors primarily employed silver-clad Bi-Sr-Ca-Cu-O (BSCCO), manufactured by the oxide powder in tube (OPIT) process [e.g. 1]. While the BSCCO/OPIT materials offer advantages in the ease of long length manufacturing of relatively stongly-linked material, their performance in magnetic fields is inherently not as good as YBCO. However, unless a high degree of crystallographic in-plane intergranular alignment is present, $YBa_2Cu_3O_{7-\delta}$ (YBCO) exhibits weak-link behavior [2,3]. This has limited YBCO to film growth on single crystal substrates, limiting its use to electronic device applications. However, the development of long lengths of flexible substrates possessing high degrees of crystallographic in-plane alignment [4,5] has led to a renewed interest in YBCO tapes for applications involving high electrical currents and magnetic fields, and a second generation of high temperature superconducting tapes.

While electrical transport measurements are the final verdict as to the quality of a given superconducting tape, x-ray diffraction (XRD) analysis possesses a great deal of predictive capability regarding the potential current-carrying capabilities of a YBCO tape. For example, the full width at half maximum of the YBCO (103) phi scan peak can in most cases predict the best J_c that one might expect from a YBCO tape [4]. Another important variable in the synthesis of YBCO is the oxygen stoichiometry, or δ in $YBa_2Cu_3O_{7-\delta}$. In bulk YBCO, this can be accurately and precisely determined by iodometric titration [6]. However, because of the minuscule amounts of material present in YBCO films, the only suitable methods available are Raman spectroscopy [7] and XRD. The most common method used to estimate oxygen stoichiometry by XRD is by employing the YBCO c-axis dependence on oxygen stoichiometry [e.g. 8,9,10]. However, this assumes that c-axis variation is only a result of changing oxygen stoichiometry. In reality, other factors such as cation disorder [11] and other defects may also cause variations in the YBCO c-axis value. A second method involves the variation of the relative intensities of the YBCO $(00l)$ peaks [12]. Again, this involves assumptions regarding the effects of other factors on the $(00l)$ intensity ratios. The object of this work was to evaluate and compare the two methods of oxygen stoichiometry estimation in order to develop methods for routine oxygen stoichiometry determination of YBCO films produced by our coated conductor program. Peak breadth analysis of $(00l)$ peaks was also performed to provide an estimate of crystalline perfection.

Films of YBCO were deposited by pulsed laser deposition at fast and slow deposition rates, and the as-deposited films were analyzed by XRD. We also analyzed high and low deposition rate films that had received a post-depositional high temperature anneal. It has been shown that while increasing deposition rate results in degradation of J_c, a high temperature post-depositional anneal of the high deposition rate films can cause recovery of some of this J_c loss [13]. Presumably, the low deposition rate films are more highly ordered and defect free than the high deposition rate films. The high temperature anneal of the high deposition rate film should eliminate at least some defects such as cation disorder and vacancies. In addition to the in situ and optional post anneal heat treatments, films received a final 823 K oxygen anneal to decrease contact resistance of silver pads that were deposited for electrical transport measurements. Two of the low deposition rate films were x-rayed before and after this low temperature anneal to evaluate any effects of this final step. The first set of analyses were performed on micro-bridge samples that had previously been used for J_c measurements (series A). A second set of unpatterned, fresh samples (series B) was also analyzed to ensure that the J_c measurement had not degraded the films prior to XRD analysis, and to examine the effect of a higher temperature post anneal. The high temperature anneal was 1063 K for 30 minutes for series A and 1098 K for 30 minutes for series B. The time/temperature processing schedule for each film is shown schematically in figure 1.

EXPERIMENTAL

The YBCO films were deposited at low (0.2-0.6 nm/s) and high (20-24 nm/s) deposition rates. The films were deposited on (100) surfaces of single crystal $SrTiO_3$ substrates using a pulsed excimer laser (KrF, wavelength=248 nm, pulse-width=25 ns, energy density=4 J/cm^2) firing on a target of stoichiometric $YBa_{2.0}Cu_{3.0}O_{7.0-\delta}$. The substrate temperature was held at 1063 K during deposition, and the distance from the target to the substrate was 4 cm. The deposition system operating pressure was 33.33 Pa (oxygen). After deposition, the chamber oxygen pressure was increased to 33,330 Pa and the sample was cooled to room temperature. Under these conditions, crystalline films on the order of 1 μm thick and cm square were obtained. One each of the high rate and low rate films

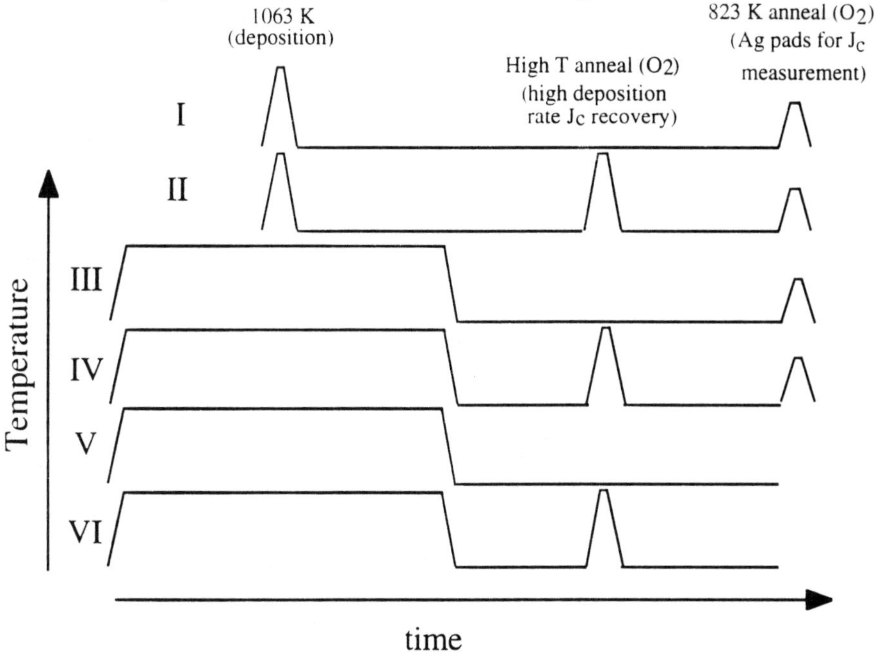

Figure 1. Schematic time/temperature processing schedule showing each of the six types of sample that were analyzed.

was subsequently annealed at 1063 or 1098 K for 3,600 seconds. All four films received a low temperature oxygen anneal at 823 K for 1,800 s.

X-ray diffraction patterns were collected from each film using a Scintag XDS2000 theta-theta powder diffractometer with a goniometer radius of 250 mm. The diffractometer was equipped with a Cu anode sealed tube x-ray source, a graphite monochromator, and a scintillation detector. Intensity data was corrected for detector deadtime using a software correction. Profile fitting was done on all peaks using a split Pearson VII profile shape function [14]. Instrumental broadening was accounted for using the NIST line profile standard LaB_6 [15]. Lattice parameters were obtained by performing a cos(θ)cot(θ) extrapolation [16]. Crystallite size and nonuniform strain values were obtained using the Gauss-squared integral breadths method [16]. The calibration curve for oxygen stoichiometry as a function of c-axis lattice parameter was obtained by fitting a 3rd-order polynomial to oxygen stoichiometry and lattice parameter data obtained from the neutron diffraction studies of bulk YBCO performed by Jorgensen, et al. (1990)[9] and Andersen, et al. (1990)[10]. Calibration curves for oxygen stoichiometry as a function of YBCO (00*l*) intensity ratios were calculated from oxygen stoichiometries and crystal structure data from Jorgensen, et al. (1990)[9] using the program GSAS [17]. Linear regression analyses were done using the spreadsheet software, Microsoft Excel, which provided estimates of line slopes as well as their 1σ standard deviations. The ±1σ standard deviations in these slopes and intercepts were propagated through the calculations to obtain uncertainties for the lattice parameters, crystallite size and nonuniform strain values.Transport critical current densities were measured on patterned microbridges 500 μm wide by 5 mm long.

RESULTS AND DISCUSSION

Figure 2 shows the variation of the two oxygen stoichiometry-dependent variables as a function of processing conditions. The c-axis lattice parameter was found to increase with increased deposition rate in the samples that received no high temperature post anneal, but returned to near the low-rate value with the high temperature post-anneal. This recovery was very pronounced in the sample that received the higher temperature of the two post anneals, with a c-axis value actually smaller than that of the low-rate value. The high temperature anneal had little or no effect on the c-axis lattice parameter values of the low-rate film. Also, the final, 823 K post-anneal had little or no effect on the c-axis lattice parameter values for the low rate films. Critical current density values are also shown in figure 2 for the microbridge samples. It can be seen that Jc tracks relative changes in the c-axis lattice parameter value quite well.

The I(004)/I(005) intensity ratio gives oxygen stoichiometry values that are divergent from those obtained from the c-axis lattice parameters, with the exception of the high-rate, post-annealed sample. Ye and Nakamura [12] have shown that cation (Y/Ba) disorder can lead to anomalously high I(004)/I(005) oxygen stoichiometry values. However, for our samples, the anomalously high oxygen stoichiometry values are found in those samples that we expected to have the highest degree of cation order. The fact that the I(004)/I(005) values for the low rate samples that did not receive the final 823 K post anneal fall very close to those that did indicates that the 823 K post anneal had no effect on the I(004)/I(005) ratios. This also indicates that there is a fairly high degree of precision associated with these ratios, and that the observed differences are probably real. It is possible that other, unaccounted for structural variables may be affecting the I(004)/I(005) ratio values in our samples. In any case, the I(004)/I(005) values in our samples (0.085 maximum) are at the lower range of those reported by Ye and Nakamura [12] for Y/Ba-disordered YBCO films (.075-0.165), so Y/Ba disorder should still be minimal in our samples. We interpret our I(004)/I(005) results as indicative of essentially fully oxygenated YBCO films possessing varying degrees of structural defects. These defects manifest themselves as anomalously high I(004)/I(005) ratio values and c-axis lattice parameter values.

The results of the peak breadth analysis of $(00l)$ peaks is shown in figure 3. While the final 823 K oxygen anneal had no effect on the c-axis lattice parameter and I(004)/I(005) ratio values, effects are seen in the crystallite thickness and nonuniform strain values that were obtained from the peak breadth analysis. Low deposition rate samples that experienced the 823 K anneal were found to have

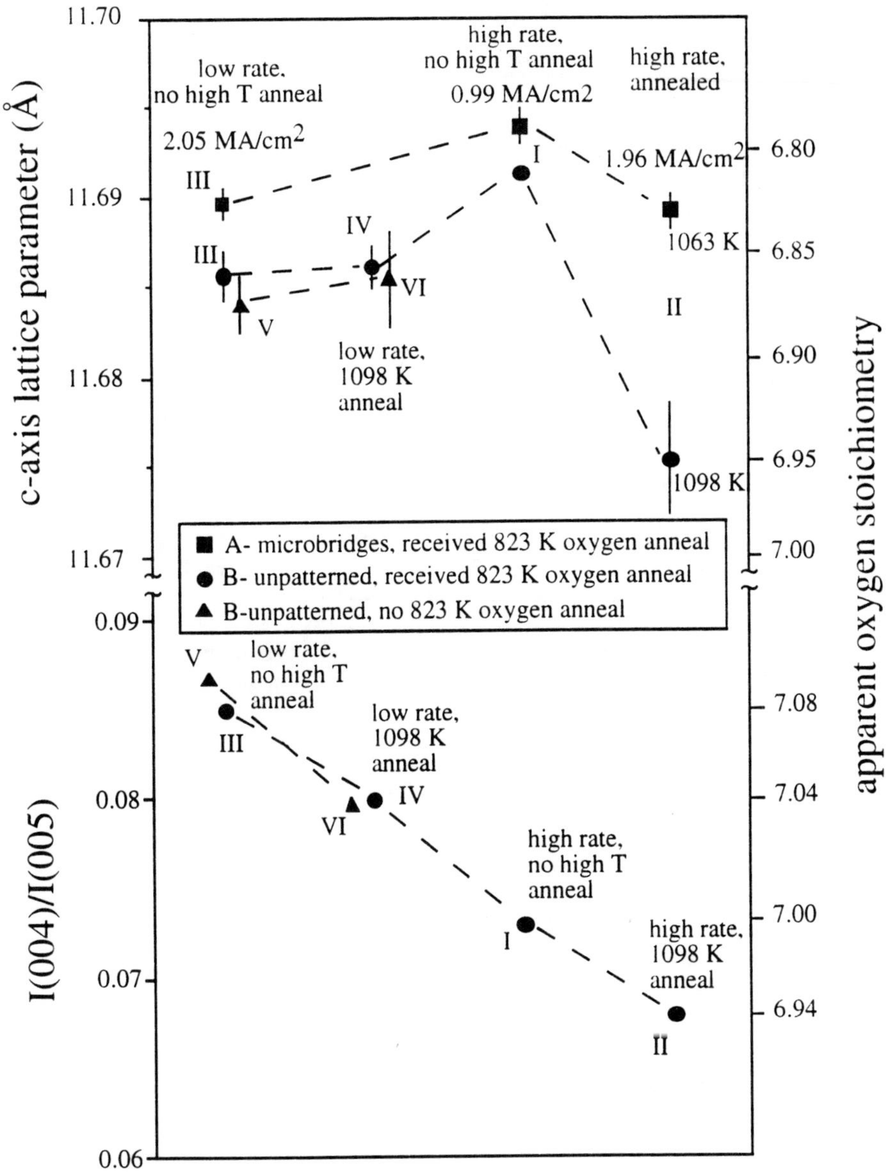

Figure 2. Variation of YBCO c-axis lattice parameter, I(004)/I(005) ratio and apparent oxygen stoichiometry with deposition rate and high temperature anneal. Dashed lines are a guide for the eye.

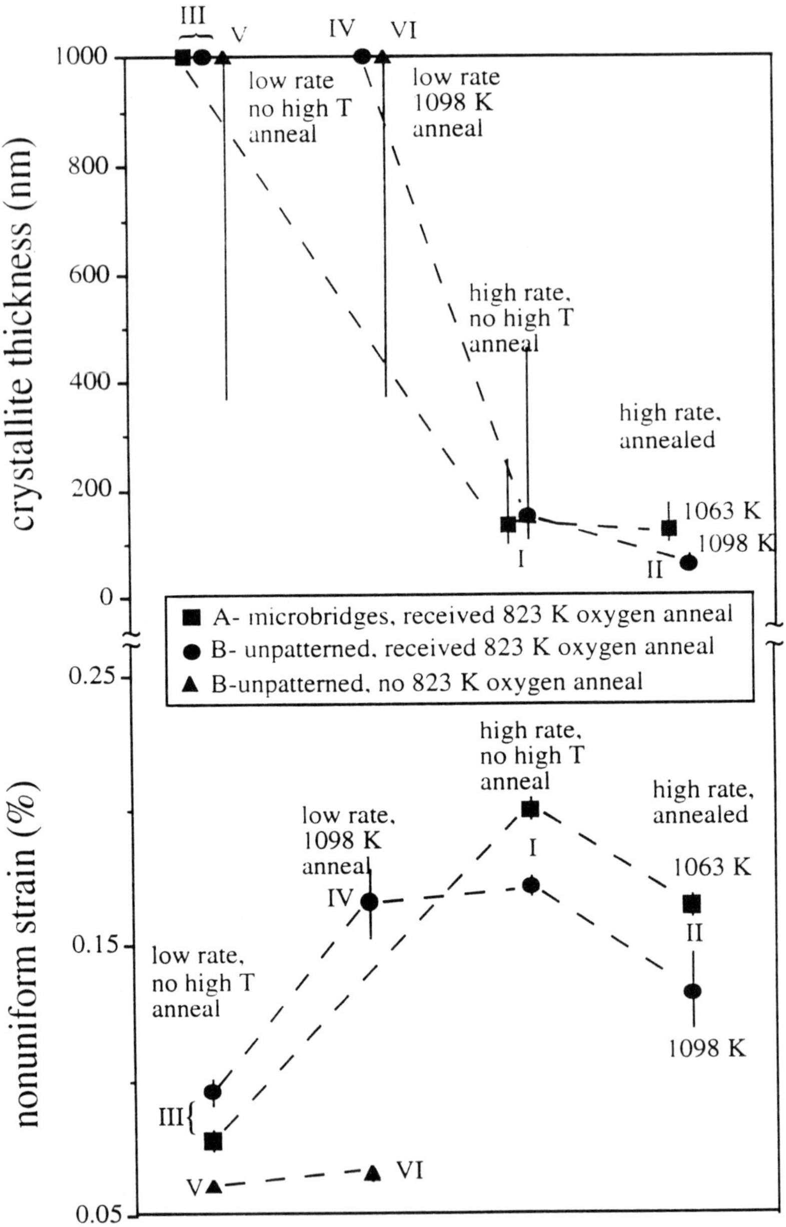

Figure 3. Variation of crystallite size and strain with deposition rate and anneal. Dashed lines are a guide for the eye.

infinitely thick (taken to be the nominal 1 micrometer film thickness) crystallites, including the lower and upper thickness value estimates based on the error estimates from the linear regression analysis. The low rate films that did not experience this anneal have low range thickness values near 400 nm, regardless of whether they experienced the high temperature anneal. The effect of the 823 K anneal is also seen in the strain values, with samples that did not experience the final anneal exhibiting significantly lower strain values than those that did. The significant influence of the relatively low temperature anneal on strain values, but not the c-axis and I(004)/I(005) ratio values suggests that this final anneal is not changing oxygen content or cation order; changes in the nature of the a-b twins may be a cause of the observed strain values in the low rate films.

Samples deposited at high deposition rates showed finite crystallite thickness values (~100 nm) independent of post anneal, indicating that crystallite thickness is frozen in at the time of deposition. The high rate films are probably composed of stacks of multiple crystallites whereas the low rate films are more monolithic in nature.

As with the c-axis lattice parameter, strain values increase with increased deposition rate, and can be reduced by a high temperature anneal, and also track J_c values obtained from the microbridge samples.

CONCLUSIONS

It has been found that while the c-axis lattice parameter may be suspect when trying to obtain accurate oxygen stoichiometry values, relative changes in c-axis lattice parameter values track relative changes in critical current density quite well. Nonuniform strain values obtained from XRD peak breadth analysis have also been found to track changes in J_c caused by varying processing conditions. Intensity ratios (I(400)/I(500)) indicate that all of the films in this study were fully oxygenated suggesting that variation in c-axis lattice parameter values, nonuniform strain values, and Jc values are not the result of oxygen variation, but are due to variations in other structural parameters.

REFERENCES

[1]W.L. Carter, G.N. Riley Jr., A. Otto, D.R. Parker, C.J. Christopherson, L.J. Masur, and D. Buczec, "Advances in the Development of Silver-Sheathed (Bi,Pb)2223 Composite Conductors", *IEEE Transactions in Applied Superconductivity*, **5** 1145-1149 (1995).

[2]D. Dimos, P. Chaudhari, J. Mannhart, and F.K. LeGoues, "Orientation Dependence of Grain Boundary Critical Currents in $YBa_2Cu_3O_{7-\delta}$ Bicrystals", *Physical Review Letters*, **61** 219-222 (1988).

[3]D. Dimos, P. Chaudhari, and J. Mannhart, "SuperconductingTransport Properties of Grain Boundaries in $YBa_2Cu_3O_{7-\delta}$ Bicrystals", *Physical Review B*, **41** 4038-4049.

[4]X.D. Wu, S.R. Foltyn, P.N. Arendt, W.R. Blumenthal, I.H. Campbell, J.D. Cotton, J.Y. Coulter, W.L. Hults, M.P. Maley, H.F. Safar, and J.L. Smith, "Properties of $YBa_2Cu_3O_{7-\delta}$ thick films on flexible buffered metallic substrates", *Applied Physics Letters*, **67** 2397-2399 (1995).

[5]A. Goyal, D.P. Norton, J.D. Budai, M. Parenthaman, E.D. Specht, D.M. Kroeger, D.K. Christen, Q. He, B. Saffian, F.A. List, D.F. Lee, P.M. Martin, C.E. Klabunde, E. Hartfield, and V.K. Sikka, "High Critical Current Density Superconducting Tapes By Epitaxial Deposition of $YBa_2Cu_3O_x$ Thick Films on Biaxially Textured Metals", *Applied Physics Letters* **69** 1795-1797 (1996).

[6]E.H. Appelman, L.R. Morss, A.M. Kini, U. Geiser, A. Umezawa, G.W. Crabtree, and K.D. Carlson, "Oxygen Content of Superconducting Perovskites $La_{2-x}Sr_xCuO_y$ and $YBa_2Cu_3O_y$", *Inorganic Chemistry* **26** 3237-3239 (1987).

[7]O. Martinez, J. Jimenez, P. Martin, A.C. Prieto, S. Degoy, D. Chambonnet, C. Belouet, S. Nicoletti, and L. Correra, "A MicroRaman Study of the Structural Properties of PLD High T_c Superconducting Thin Films", *Physica C* **270** 144-154 (1996).

[8]R.J. Cava, B. Batlogg, C.H. Chen, E.A. Rietman, S.M. Zahurak, and D. Werder, "Single-Phase 60-K Bulk Superconductor In Annealed $YBa_2Cu_3O_{7-\delta}$ $(0.3<\delta<0.4)$ With Correlated Oxygen Vacancies in the Cu-O chains", *Physical Review B* **36** 5719-5722.

[9]J.D. Jorgensen, B.W. Veal, A.P. Paulikas, L.J. Nowicki, G.W. Crabtree, H. Claus, and W.K. Kwok,"Structural Properties of Oxygen-Deficient $YBa_2Cu_3O_{7-\delta}$", *Physical Review B* **41**,1863-1877 (1990).

[10]N.H. Andersen, B. Lebech, and H.F. Poulsen, "The Structural Phase Diagram and Oxygen Partial Pressure of $YBa_2Cu_3O_{6+x}$ Studied by Neutron Powder Diffraction and Gas Volumetry", *Physica C* **172** 31-42 (1990).

[11]J. Ye and K Nakamura, "Systematic Study of the Growth-Temperature Dependence of Structural Disorder and Superconductivity in $YBa_2Cu_3O_{7-\delta}$ Thin Films", *Physical Review B* **50**, 7099-7106 (1994).

[12]J. Ye and K Nakamura, "Quantitative Structure Analyses of $YBa_2Cu_3O_{7-\delta}$ Thin Films: Determination of Oxygen Content From X-ray Diffraction", *Physical Review B* **48**, 7554-7564 (1993).

[13] S.R. Foltyn, E.J. Peterson, J.Y. Coulter, P.N. Arendt, Q.X. Jia, P.C. Dowden, M.P. Maley, X.D. Wu, and D.E. Peterson, "Influence of Deposition Rate on the Properties of Thick $YBa_2Cu_3O_{7-\delta}$ Films Produced by Pulsed Laser Deposition", Submitted to *Journal of Materials Research* (1997).

[14]S.A. Howard and K.D. Preston, "Profile Fitting of Powder Diffraction Patterns"; pp. 217-272 in *Modern Powder Diffraction* Edited by D.L. Bish and J.E. Post. Mineralogical Society of America, Washington D.C., 1989.

[15]T.G. Fawcett, C.E. Crowder, S.J. Brownell, Y. Zhang, C.R. Hubbard, W. Schreiner, C.P. Hamill, T.C. Huang, E. Sabino, J.I. Langford, R. Hamilton, and D. Louer, "Establishing an Instrumental Peak Profiling Calibration Standard for Powder Diffraction Analysis: International Round Robin Conducted by the JCPDS-ICDD and the U.S. National Bureau of Standards", *Powder Diffraction* **3**, 209-218 (1988).

[16]H.P. Klug and L.E. Alexander, *X-ray Diffraction Procedures for Polycrystalline and Amorphous Materials*, 2nd Edition, John Wiley and Sons, New York (1974).

[17]A.C. Larson and R.B. Von Dreele, "GSAS-Generalized Crystal Structure Analysis System", Los Alamos National Laboratory Report LA-UR-86-748 (1987).

FORMATION OF $Bi_2Sr_2CaCu_2O_8$ (Bi-2212) FROM METAL-BEARING PRECURSORS

Terry J. Detrie, Ken H. Sandhage
Department of Materials Science and Engineering
The Ohio State University, Columbus, OH 43210

ABSTRACT

A powder-metallurgical method for producing Ag-sheathed $Bi_2Sr_2CaCu_2O_8$ (Bi-2212) tapes from Bi_2O_3-Sr-Ca-Cu-bearing precursors is presented. Sr, Ca, and Cu are relatively ductile metals that can endow a precursor with desired plasticity for deformation processing. Bi_2O_3 was used, instead of metallic Bi, to enhance powder yields upon high-energy mechanical alloying and to avoid low-melting, metallic liquids during subsequent heat treatment. Milled Bi_2O_3-Sr-Ca-Cu powder mixtures were packed into a silver tube and then drawn and rolled into tapes. A series of isothermal annealing treatments at 300-400°C, 750°C, and then 860°C were used to oxidize the metallic phases in the core of the tape and to allow for conversion into Bi-2212. The oxidation kinetics, microstructural features, and the reaction sequence leading to Bi-2212 are discussed.

INTRODUCTION

A standard process for fabricating long lengths of BSCCO superconducting wires and tapes is the oxide-powder-in-tube (OPIT) method [1-8]. In this process, silver-sheathed $Bi_2Sr_2CaCu_2O_8$ (Bi-2212) or $(Bi,Pb)_2Sr_2Ca_2Cu_3O_{10}$ (Bi-2223) conductors are produced by packing oxide powder of desired stoichiometry into a silver tube, which is then deformed (e.g., by extrusion, drawing, rolling, pressing, or combinations thereof) into a wire or tape geometry. BSCCO/Ag composites with multilaminate or multifilamentary geometries can be produced by co-deformation of multiple, silver-sheathed rods or tapes [1-5]. For Bi-2212 conductors, a final annealing treatment is usually conducted at a peak temperature above the incongruent melting point of Bi-2212 [4-8]. Slow cooling through, and prolonged annealing below, the oxide solidus temperature can lead to a high degree of Bi-2212 grain alignment and a high transport critical current density ($J_c \geq 100$ kA/cm^2 at 4.2K, 0 applied field, 1μV/cm criterion), particularly for silver-sheathed tapes with thin (usually ≥40 μm) oxide cores of uniform thickness [4,5,7,8].

An alternate method for fabricating long lengths of silver-sheathed BSCCO wire or tape is the metal-powder-in-tube (MPIT) technique [9-12]. In this process, mechanically-alloyed powder, consisting of an intimate mixture of elements of a desired stoichiometry, is packed into a silver tube and deformed into a wire or tape as discussed above for OPIT processing. The ductility of metal-bearing mixtures allows for the fabrication of relatively thin, dense, and geometrically-uniform precursors by such deformation processing. For example, Otto, et al. [9-11] have reported the fabrication of multifilamentary Bi-Pb-Sr-Ca-Cu/Ag composite precursors with individual Bi-Pb-Sr-Ca-Cu-bearing filaments ≤ 5 microns in thickness. Selective oxidation of the non-noble elements and post-oxidation annealing (e.g., partial melt processing of Bi-2212-bearing tapes or thermo-mechanical processing of Bi-2223-bearing tapes) can then yield superconducting BSCCO filaments or layers within a noble silver matrix. An average critical current density of 15.2 kA/cm^2 at 77K (0 applied field, 1μV/cm criterion) has been reported for 50 m lengths of multifilamentary Bi-2223/Ag tapes produced in this manner [9]. Although several papers have been published on the fabrication of BSCCO/Ag composites by the MPIT process [9-11], important aspects of processing (e.g., conditions of mechanical alloying, deformation processing, oxidation and post-oxidation annealing) and microstructural/phase evolution have not been reported. The purpose of the present paper is to describe several important processing features associated with the fabrication of silver-sheathed Bi-2212 tapes from metal-bearing (Bi_2O_3-Sr-Ca-Cu) precursors.

EXPERIMENTAL PROCEDURE

A powder-metallurgical route, illustrated schematically in Fig. 1, was used to prepare Ag-sheathed Bi_2O_3-Sr-Ca-Cu precursor tapes. Precursor powders were prepared by high-energy mechanical alloying. The purities, forms, and sizes of the starting components are listed in Table I. A 10-12 gram mixture of Bi_2O_3 powder, Sr granules, and Ca granules, with a molar Bi_2O_3:Sr:Ca ratio of 1:2:1, was first

Table I - Starting Components of the Mechanically-Alloyed Precursors

Component	Form	Size (Ave. Diameter or Largest Dimension)	Purity
Bi_2O_3#	Powder	≈ 15 μm	99.9%
Sr*	Granules	≤ 6 mm	99.8%
Ca†	Granules	≤ 6 mm	99.5%
Cu†	Powder	< 89 μm	99.5%

#Aldrich Chemical Co., Milwaukee, WI *ProChem, Inc., Rockford, IL
†Johnson Matthey, Inc., Ward Hill, MA

sealed inside a hardened steel vial (3.8 cm internal diameter x 5.7 cm internal length) along with 2 hardened steel balls (1.3 cm diameter) and $\approx$50 ml of hexane within a purified Ar atmosphere glove box. The mixture was mechanically alloyed for 4 hours within a high-energy vibratory ball mill (Model 8000 Mixer-Mill, SPEX Industries, Edison, NJ). Cu powder was then added to the milling vial (to obtain a Bi_2O_3:Sr:Ca:Cu molar ratio of 1:2:1:2) within the glove box and milling was conducted for an additional 0.5 to 4 hours. Hexane was removed from the milled powders by filtration and evaporation in the Ar atmosphere glove box.

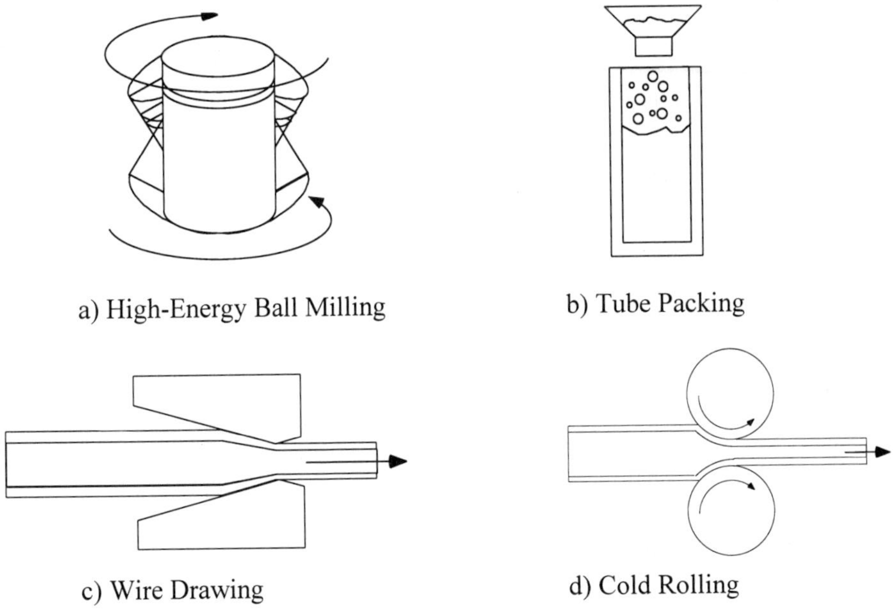

Fig. 1. Schematic illustration of the powder-metallurgical route used to prepare silver-sheathed Bi_2O_3-Sr-Ca-Cu tapes.

To allow for reproducible processing, all powder handling operations (e.g., mixing, milling, tube packing) were conducted within an argon atmosphere glove box. The argon was purified by passage through a powder bed of Drierite (W. A. Hammon Drierite Co., Xenia, OH) and Ascarite II™ (Thomas Scientific, Swedesboro, NJ) to remove residual water vapor and carbon dioxide, as well as through a Model MO-5 Dri-Train gas purification system (Vacuum Atmospheres Co., Hawthorne, CA) to remove residual oxygen. The oxygen content of the glove box atmosphere was maintained at ≤ 1 ppm, as indicated by a zirconia oxygen sensor (Model TM-1B, Ametek, Inc., Pittsburgh, PA).

The milled powder was packed into a 15 cm long X 10.3 mm OD silver tube (wall thickness = 1 mm, 99.99% purity, Refining Systems, Las Vegas, NV). The packed tube was evacuated, welded shut, and then drawn through a series of dies into a 3.5 mm diameter wire. Drawing was conducted at an average speed of 6 cm/sec with ≤ 8% reduction per pass. The drawn wire was rolled into a tape with a two-high rolling mill (8.5 cm diameter rolls) operating at a rolling speed of 25 cm/sec with ≤ 9% reduction per pass. The final tapes possessed continuous precursor cores with average thicknesses ranging from 150 μm up to 550 μm.

The Ag-sheathed Bi_2O_3-Sr-Ca-Cu tapes were cut into 10 mm lengths for subsequent oxidation processing. Oxidation and post-oxidation annealing were conducted in flowing $O_2(g)$ according to the schedule shown in Fig. 2. Prior to entry into the reaction tube of the furnace, the oxygen was passed through a packed bed of Drierite and Ascarite II™ to remove residual water vapor and carbon dioxide. To avoid the back-flow of air into the reaction tube, O_2(g) was bubbled through a low vapor pressure oil bath (Fisher Brand 19, Inland Vacuum Industries, Churchville, NY) located downstream of the reaction tube.

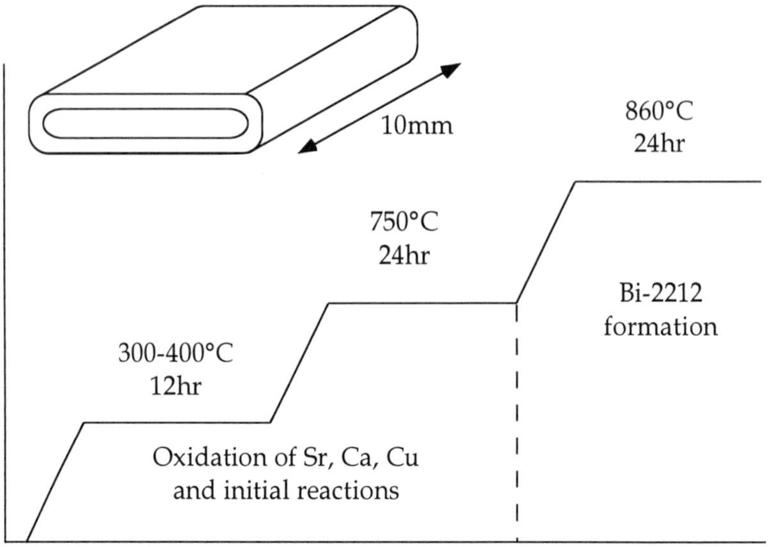

Fig. 2. Heat treatment schedule used to transform the Bi_2O_3-Sr-Ca-Cu precursors.

Precursor and transformed specimens were characterized by x-ray powder diffraction (XRD) analyses, optical microscopy, and electron microscopy (SEM). The silver-sheathed specimens were transferred inside the Ar atmosphere glovebox, whereupon the silver sheath was stripped from the core. The core of the tape was then ground into powder with a mortar and pestle, and mixed with silver powder (an internal XRD standard) and silicone grease (high vacuum silicone grease, Dow Corning, Midland, MI). The powder-grease mixture was spread onto a single crystal silicon substrate that had been ground to expose an irrational plane, so as not to generate diffraction peaks within the 2θ range examined. XRD analyses were conducted at room temperature with Cu-Ka radiation at a scan rate of 1°/min. Specimens were also mounted in epoxy, cross-sectioned with a diamond wafering blade, and polished with diamond pastes in kerosene down to a 1 micron finish. Electron microscopy was conducted with the use of a Philips XL-30 scanning electron microscope (Philips Electronics N.V., Eindhoven, The Netherlands) equipped with a field emission gun and an EDS Si/Li detector (Edax International, Mahwah, NJ). Image analyses were conducted with the use of Buehler Omnimet 4 software (Clemex Technologies, Inc., Longueuil, Canada). Thermogravimetric (TG) analyses (Model TG-171, Cahn Instruments, Inc., Cerritos, CA) were also conducted to evaluate oxidation kinetics in flowing, purified oxygen.

RESULTS AND DISCUSSION

Processing/Characterization of Silver-Sheathed Precursor Tapes

Initial attempts to mechanically alloy an all-metallic Bi-Sr-Ca-Cu precursor for times of 4-8 hours were unsuccessful, owing to appreciable agglomeration and welding of the powder to the milling media and vial. Such agglomeration and welding were avoided by replacing metallic Bi with Bi_2O_3 and conducting the milling treatment in two stages. In the first stage, Bi_2O_3 powder was milled with Sr and Ca granules for 4 hours. Powder yields of ≥ 98% were obtained from the milling vial after this treatment. After the addition of copper powder to obtain a precursor stoichiometry of Bi_2O_3:Sr:Ca:Cu = 1:2:1:2, a second milling treatment was conducted for 0.5 to 4 hours. As shown in Fig. 3, milling of the copper-bearing mixture for 4 hours resulted in a much more refined microstructure than milling for 0.5 hours. The largest phases detected in specimens of the type shown in Fig. 3a (4 hours of second stage milling) possessed diameters < 10 microns, whereas elongated copper grains with lengths of ≥ 50 microns were observed in specimens of the type shown in Fig. 3b (0.5 hours of second stage milling). Powder yields of ≥ 96% were obtained after completion of the two-stage (4 hours + 4 hours) milling treatment. EDS analyses of the core of a precursor tape of the type shown in Fig. 3b yielded an average Bi:Sr:Ca:Cu molar ratio of 2.0:2.1:1.0:1.9 (18 analyses with each analysis conducted over an area of 0.5 mm^2, normalized to Bi = 2.0), which was not far from the targeted stoichiometry.

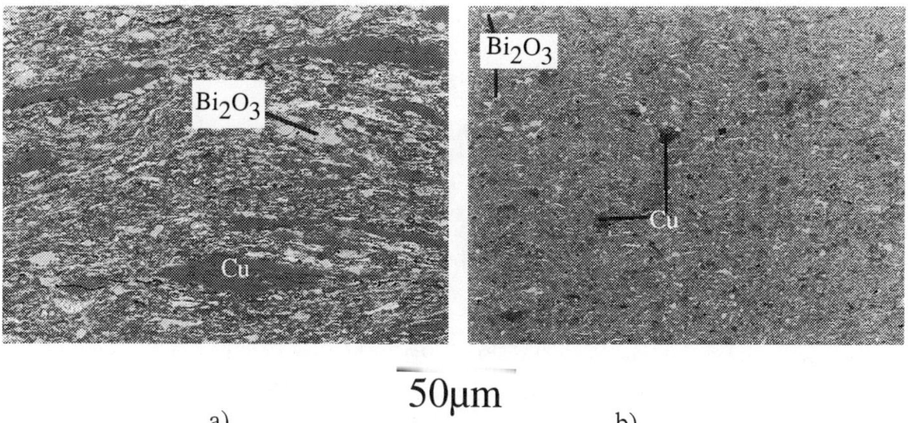

Fig. 3. Backscattered electron images of the cores of Ag-sheathed precursor tapes. Precursor powder was produced with two-stage mechanical alloying. In the first stage, a Bi_2O_3-Sr-Ca mixture was milled for 4 hours. Cu powder was then added and a milling was conducted for: a) 0.5 or b) 4 more hours. The grey matrix phase is Sr and Ca.

One concern with high-energy milling was a reduction in the volume fraction of ductile metal in (and, hence, the formability of) the precursor due to the formation of brittle intermetallic compounds. Several binary intermetallic compounds are known to exist within the Bi-Sr, Bi-Ca, Sr-Cu, and Ca-Cu systems [13]. Predominant diffraction peaks for binary intermetallic compounds (e.g., Bi_3Sr_5, $BiSr_2$, $Bi_{10}Ca_{11}$, Bi_3Ca_5, $CaCu_5$, Ca_2Cu, or CaCu) were not detected, however, after the two-stage (4 hours + 4 hours) milling treatment.

Drawing and rolling of the silver-sheathed precursor powder yielded dense precursor tapes. Image analyses of polished cross-sections of silver-sheathed tapes of the types shown in Figs. 3a and 3b indicated that the precursor cores of such tapes were ≥ 99% dense. Compaction to such a high green density was not surprising in light of the ductile metal content of the precursors. A Bi-2212 precursor mixture of Bi_2O_3-Sr-Ca-Cu consists of 68.5 vol% ductile metal (i.e., Sr+Ca+Cu).

Oxidation and Oxide Phase Evolution

Initial isothermal anneals were conducted at 300-400°C to allow for the oxidation of solid Ca and Sr, so as to avoid degradation of the tape geometry due to the formation of low-melting Ag-Sr or Ag-Ca eutectic liquids (the lowest solidus temperatures in the binary Ag-Sr and Ag-Ca systems are 436°C and 470°C, respectively [13]). Complete oxidation of the Sr and Ca (to form SrO and CaO) within the silver-sheathed Bi_2O_3-Sr-Ca-Cu tapes should have resulted in a weight gain of ≈ 3.0%. TG analyses conducted in pure, flowing $O_2(g)$ revealed that such a weight gain was achieved within 9 hours at 350°C and within 4 hours at 400°C.

Backscattered electron images, and associated x-ray dot maps, of longitudinal cross-sections of precursor tapes after oxidation for 12 hours at 350°C and 8 hours at 400°C are shown in Figs. 4 and 5, respectively (note: the powder within these precursors was produced with the use of a 0.5 hour second-stage milling treatment). After oxidation at 400°C, a bright Bi-rich phase was detected within the Ag sheath. Continuous Bi-rich layers were also present at the Ag/precursor interface and at the external Ag surface. Much less bismuth was found in the Ag sheath after oxidation at 350°C or 300°C. The loss of Bi from the precursor core to the sheath was attributed to the formation and outward migration of a metallic, Bi-rich liquid. Metallic Bi melts at 271.4°C [13] and could have been produced within the precursor tapes by the following oxidation/reduction reactions:

$$Bi_2O_3(s) + 3Sr(s) = 3SrO(s) + 2Bi(l) \quad (1)$$

$$Bi_2O_3(s) + 3Ca(s) = 3CaO(s) + 2Bi(l) \quad (2)$$

Thermodynamic data indicate that these reactions are strongly favored at 300-400°C (e.g., the standard Gibbs free energies of reactions (1) and (2) are -1,180 kJ/mole and -1,300 kJ/mole, respectively, at 400°C [14]). Hence, the oxidation of strontium and calcium during heat treatment at 300-400°C could have occurred either by reactions (1) and (2) or by reaction of these metals with gaseous oxygen. The reduction in the amount of bismuth present within the silver sheath, upon lowering the initial isothermal oxidation temperature from 400°C to 350°C, is consistent with a reduction in the extent of strontium and calcium oxidation by reactions (1) and (2) relative to oxidation by reaction with gaseous oxygen; that is, the decrease in the

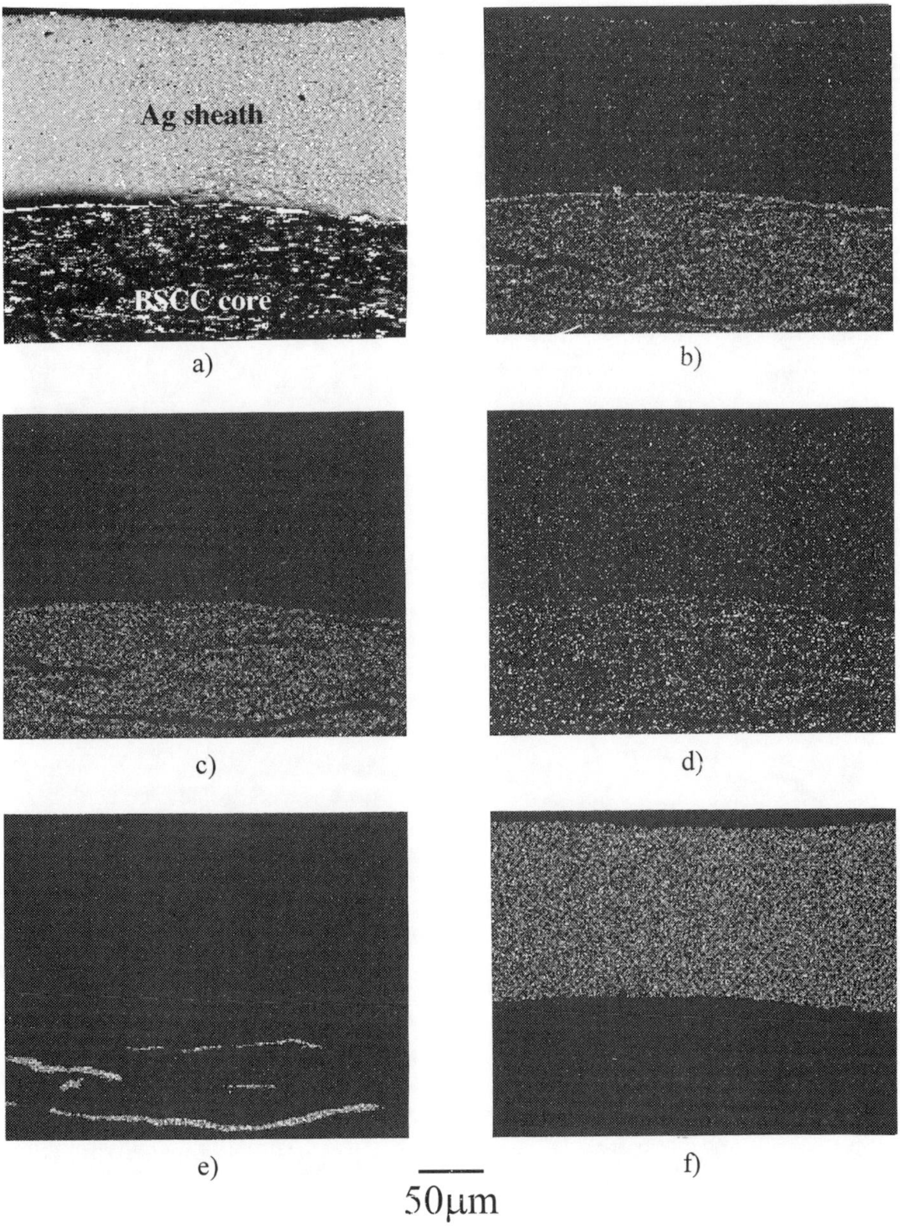

Fig. 4. a) A backscattered electron image b)-f) associated x-ray maps for Bi, Sr, Ca, Cu, and Ag, respectively, of a longitudinal cross-section of a silver-sheathed precursor after oxidation for 12 hours at 350°C.

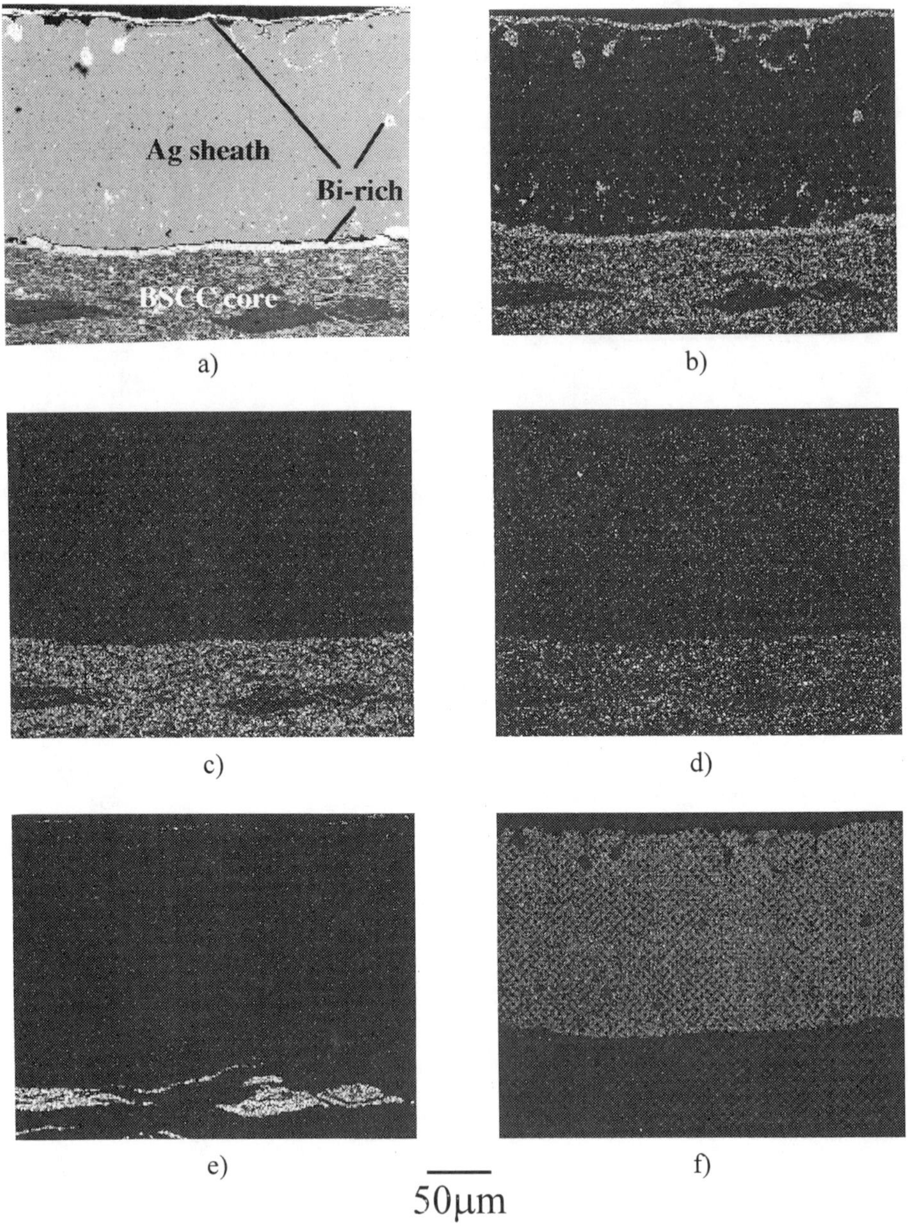

Fig. 5. a) A backscattered electron image b)-f) associated x-ray maps for Bi, Sr, Ca, Cu, and Ag, respectively, of a longitudinal cross-section of a silver-sheathed precursor after oxidation for 8 hours at 400°C.

rate of condensed-phase oxidation of Sr and Ca from 400 to 350°C was apparently greater than the decrease in the rate of gas-phase oxidation. In any event, an initial oxidation temperature of 350°C was chosen for subsequent specimen processing in order to minimize the loss of bismuth to the silver sheath.

A second isothermal anneal was conducted at 750°C for 24 hours in pure, flowing oxygen to allow for: 1) the oxidation of residual copper at a temperature below the 780°C eutectic of the Ag-Cu system [13], and 2) evaluation of the intermediate oxide compounds produced by reaction of the resulting oxide mixture. XRD patterns obtained from the precursor core of silver-sheathed tapes after isothermal anneals at 350°C (12 hours), 750°C (24 hours), and 860°C (24 hours) are shown in Fig. 6 (note: the precursor powder for the tapes associated with Fig. 6 was produced with a second stage milling treatment of 4 hours).

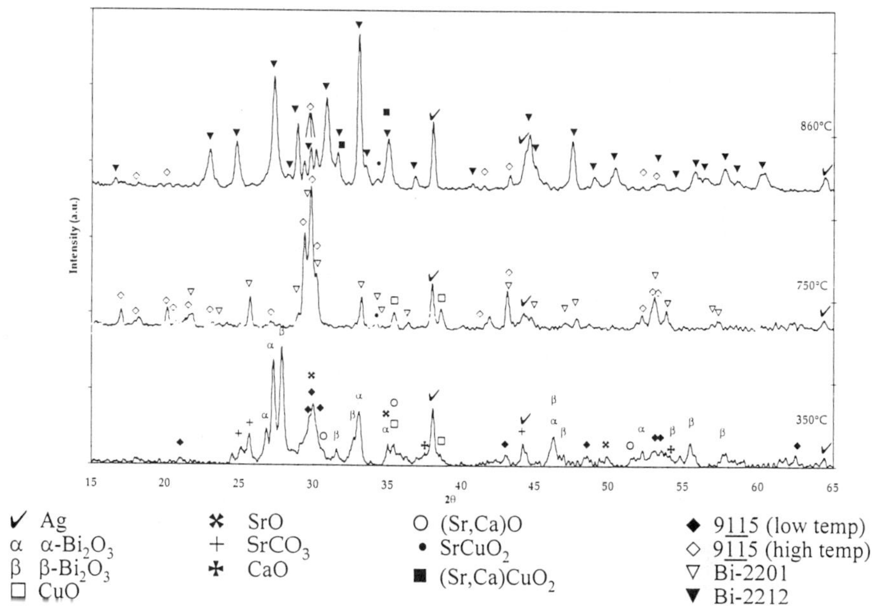

Fig. 6. XRD patterns obtained from the precursor tapes after heat treatment for a) 12 hours at 350°C, b) 24 hours at 750°C, and c) 24 hours at 860°C.

The presence of distinct diffraction peaks for Bi_2O_3 (both monoclinic and tetragonal polymorphs), CuO, and $SrCO_3$, and modest or weak peaks for SrO, CaO, and (Sr,Ca)O, along with the absence of peaks for metallic Bi, Sr, Ca, or Cu, indicated that the precursor of Fig. 6a had been completely oxidized within 12 hours at 350°C. (The formation of strontium carbonate was attributed to the reaction of oxidized strontium with carbon dioxide in ambient air during transfer of the powder specimen from the glove box to the diffractometer.) Diffraction peaks consistent with a low-temperature, orthorhombic polymorph of a copper-free, $Bi_9Sr_{11}Ca_5O_y$-type (Bi-9<u>11</u>50) phase [15-18] were also detected after the 350°C heat treatment. The formation of this ternary oxide compound during such a modest annealing treatment was somewhat surprising, in that the fabrication of this copper-free phase from mixed salts (e.g., a mixture of Bi_2O_3, $SrCO_3$, and $CaCO_3$) is usually conducted by heat treatment at temperatures ≥ 800°C [15-18]. In the present process, the formation of strontium and calcium carbonate were avoided during heat treatment at ≥ 350°C, owing to the absence of carbonaceous material in the precursor (e.g., organic binders, plasticisers, etc.) and of CO_2(g) in the oxidizing gas stream. Hence, the formation of the Bi-9<u>11</u>50 phase could occur by the direct reaction of the oxides (i.e., without the need for $SrCO_3$ or $CaCO_3$ decomposition in advance of such reaction[1]). Prior work with other mechanically-alloyed, alkaline-earth-metal-bearing precursors has also shown that binary or ternary oxide compounds (e.g., alkaline earth cerates, titanates, ferrites, aluminates, and silicates) can be produced at considerably lower temperatures than are required for syntheses involving carbonate-bearing precursors [12,21-24].

Further reaction of the oxidized precursor for 24 hours at 750°C resulted in complete consumption of Bi_2O_3 and yielded a mixture of a monoclinic Bi-9<u>11</u>50 phase (the high-temperature polymorph that is stable at ≥ 640°C in pure oxygen [15]), the $Bi_2(Sr,Ca)_2Cu_1O_6$-type (Bi-2201) phase, and CuO. A weight loss of ≈1% was detected after the 24 hour anneal at 750°C, which was consistent with the conversion of Bi^{+5} within orthorhombic Bi-9<u>11</u>50 [15-18] into Bi^{+3} within monoclinic Bi-9<u>11</u>50 or within Bi-2201. Bi-2201 was then consumed after 24 more hours of annealing at 860°C to yield Bi-2212 as the predominant phase, along with residual Bi-9<u>11</u>50 and a small amount of alkaline earth cuprate ($SrCuO_2$, $(Sr,Ca)CuO_2$). Although the Bi-9<u>11</u>50 phase has not been reported to be one of the intermediate phases produced during the conversion of Bi_2O_3-$SrCO_3$-$CaCO_3$-CuO precursors into Bi-2212, this phase has been detected during the conversion of nitrate precursors [18]. Several authors have also reported the formation of a $Bi_2Sr_{3-x}Ca_xO_z$-type phase upon selective reduction of Cu^{++} to Cu^{+} in Bi-2212 [18,25,26]. Subsequent exposure of such reduced, $Bi_2Sr_{3-x}Ca_xO_z$-bearing samples to oxidizing conditions (e.g., at 850°C in pure oxygen) has yielded Bi-9<u>11</u>50 prior to Bi-2212 formation [18].

Work is currently underway to produce Bi_2O_3-Sr-Ca-Cu precursor tapes with Bi-rich, (Sr+Ca)-poor compositions to allow for more complete formation of Bi-2212 during post-oxidation annealing. Oxidized tapes are also being exposed to partial-melt thermal processing in order to produce high-J_c, Bi-2212 conductors.

CONCLUSIONS

A powder-metallurgical method has been used to fabricate silver-sheathed, Bi_2O_3-Sr-Ca-Cu tapes (precursors to Ag-sheathed Bi-2212). High yields (≥ 96%) of intimately-mixed, metal+oxide powder were obtained via a two-stage mechanical

---

[1] Carbonate decomposition in Bi_2O_3-$SrCO_3$-$CaCO_3$-CuO precursors to Bi-2212 is reported to commence at ≥ 650-700°C during firing in air [19,20].

alloying treatment. Dense precursor tapes were produced by sealing the milled powder into a silver tube, and then drawing and rolling the packed tube into a tape. Exposure of the Ag-sheathed precursor tapes to pure, flowing oxygen at 400°C resulted in appreciable migration of bismuth into the sheath, presumably due to the formation of metallic, Bi-rich liquid (i.e., by the reaction of Bi_2O_3 with Sr and/or Ca). By reducing the initial heat treatment temperature to ≤ 350°C, complete oxidation of the strontium and calcium could be conducted without appreciable Bi migration into the sheath. Reaction of the oxide phases within the precursor tape during a 12 hour anneal in pure, flowing oxygen at 350°C yielded orthorhombic Bi-91150. Further annealing for 24 hours at 750°C resulted in the conversion of the Bi-91150 phase into a monoclinic polymorph and the formation of Bi-2201. X-ray diffraction analyses indicated that the Bi-2201 was completely consumed during a subsequent 24 hour anneal at 860°C to yield Bi-2212 as the predominant phase.

ACKNOWLEDGMENTS

Financial support from the National Science Foundation (Grant #MSS-9211281) and the Midwest Superconductivity Consortium is gratefully acknowledged.

REFERENCES

1. G. N. Riley, Jr., W. L. Carter, "Commercialization of Ceramic Superconductor Composite Wires," Bull. Am. Ceram. Soc., **72** [7] (1993) 91-96
2. P. Haldar, L. Motowidlo, "Processing High Critical Current Density Bi-2223 Wires and Tapes," J. Met., **42** [10] (1992) 54-58
3. K. Sato, T. Hikata, H. Mukai, M. Ueyama, N. Shibuta, T. Kato, T. Masuda, M. Nagata, K. Iwata, T. Mitsui, "High-J_c Silver-Sheathed Bi-Based Superconducting Wires," IEEE Trans. Magn., **27** (1991) 1231-1238
4. K.H. Sandhage, G.N. Riley, W.L. Carter, "Critical Issues in OPIT Processing of High-J_c BSCCO Superconductors," J. Met., **43** [3] 21-25 (1991)
5. E. E. Hellstrom, "Processing Bi-Based High-T_c Superconducting Tapes, Wires, and Thick Films for Conductor Applications," pp. 383-440 in High-Temperature Superconducting Materials Science and Engineering, Ed. Donglu Shi, Pergamon Press, 1995
6. K. Heine, J. Tenbrink, M. Thoner, "High Field Critical Current Densities in $Bi_2Sr_2Ca_1Cu_2O_{8+x}$/Ag-Wires," Appl. Phys. Lett., **55** [23] 2441-2443 (1989)
7. R. D. Ray, II, E. E. Hellstrom, "Microstructurally Optimized Heat Treatment for Melt-Processed, Ag Clad $Bi_2Sr_2Ca_1Cu_2O_y$ Tape: Step Solidification Melt Processing," Physica C, **251**, 27-49 (1995)
8. T. Haugan, S. Patel, M. Pitsakis, F. Wong, S. J. Chen, D. T. Shaw, "Recent Status on High Temperature Superconducting $Bi_2Sr_2Ca_1Cu_2O_{8+x}$ Wire Development at NYSIS: 1-90 Meter Length J_cs and 3 Meter Diameter Ring Furnace Design," J. Electron. Mat., **24** [12] 1811-1815 (1995)
9. A. Otto, L. J. Masur, C. Craven, D. Daly, E. R. Podtburg, J. Schreiber, "Progress Towards a Long Length Metallic Precursor Process for Multifilament Bi-2223 Composite Superconductors," IEEE Trans. Appl. Supercond., **5** [2] (1995) 1154-1157
10. A. Otto, C. Craven, D. Daly, E. R. Podtburg, J. Schreiber, L. J. Masur,

"Properties of High-Tc Wires Made by the Metallic Precursor Process," J. Met., **45** [9] (1993) 48-52

11. A. Otto, L. J. Masur, J. J. Gannon, E. R. Podtburg, D. Daly, G. J. Yurek, A. P. Malozemoff, "Multifilamentary Bi-2223 Composite Tapes Made by a Metallic Precursor Route," IEEE Appl. Supercond., **3** [1] (1993) 915-22
12. K. H. Sandhage, L. J. Masur, G. D. Smith, J. M. Poole, M. G. McKimpson, "The Metallic Precursor Approach to Long Lengths of $YBa_2Cu_3O_x$ Superconducting Wire," pp. 347-362 in Proc. Symp. High Temp. Superconducting Compounds III: Processing and Microstructure-Property Relationships, Ed. S. H. Whang, A. DasGupta, E. Collings, TMS, Warrendale, PA, 1991
13. Binary Alloy Phase Diagrams, Second Edition, Vols. 1-2, Edited by T. B. Massalski, H. Okamoto, P. R. Subramanian, L. Kacprzak, ASM, 1990
14. I. Barin, Thermochemical Data of Pure Substances, VCH Verlagsgesellschaft, Weinheim, Germany, 1989
15. C. J. Rawn, R. S. Roth, B. P. Burton, M. D. Hill, "Phase Equilibria and Crystal Chemistry in Portions of the System SrO-CaO-1/2Bi_2O_3-CuO: V, The System SrO-CaO-1/2Bi_2O_3," J. Am. Ceram. Soc., **77** [8] 2173-78 (1994)
16. D. H. Chmielewski, K. H. Sandhage, "Stability of $Bi_2Sr_2Ca_1Cu_2O_{8\pm\delta}$ Thick Films at Elevated Oxygen Pressures and Temperatures," J. Am. Ceram. Soc., **78** [9], 2504-2512 (1995)
17. Y. H. Li, J. L. MacManus-Driscoll, "Identification of the High Pressure Phase in Bi-Sr-Ca-Cu-O Superconductors by HRTEM and XRD," Physica C, **267**, 337-344 (1996)
18. A. P. Baker, B. A. Glowacki, "The Dependence of $Bi_9Sr_{11}Ca_5$ Oxide on the Stoichiometry of Bi-2212 and its Reversible Phase Transition in Oxidizing and Non-Oxidizing Annealing Conditions," Physica C, **223** (1994) 383-390
19. W. Lo, B. A. Glowacki, "A Study of the Formation Processes of the 2212 Phase in Bi-based Superconductor Systems," Physica C, **193** 253-263 (1992)
20. G. R. Paz-Pujalt, "Solid State Reactions in the Formation of Bi-Sr-Ca-Cu-Oxide Superconductor from Metallo-Organic Precursors," Physica C, **166**, 177-184 (1990)
21. H. J. Schmutzler, K. H. Sandhage, J. C. Nava, "The Fabrication of Dense, Shaped Barium Cerate by the Oxidation of Solid Metal-Bearing Precursors," J. Am. Ceram. Soc., **79** [6], 1575-1584 (1996)
22. H. J. Schmutzler, M. M. Antony, K. H. Sandhage, "A Novel Reaction Path to $BaTiO_3$ by the Oxidation of a Solid Metallic Precursor," J. Am. Ceram. Soc., **77** [3], 721-729 (1994)
23. G. A. Ward, K. H. Sandhage, "Synthesis of Barium Hexaferrite by Oxidation of a Metallic Ba-Fe Precursor," J. Am. Ceram. Soc., **80** [6] 1508-16 (1997)
24 H. J. Schmutzler, K. H. Sandhage, "Transformation of Ba-Al-Si Precursors to Celsian (monoclinic $BaO{\cdot}Al_2O_3{\cdot}2SiO_2$) by High-Temperature Oxidation," Met. Trans. B., **26B** [1], 135-148 (1995)
25. L. M. Rubin, T. P. Orlando, J. B. Vander Sande, G. Gorman, R. Savoy, R. Swope, R. Beyers, "Phase Stability Limits of $Bi_2Sr_2Ca_1Cu_2O_{8+\delta}$ and $Bi_2Sr_2Ca_2Cu_3O_{10+\delta}$," Appl. Phys. Lett., **61** [16] 1977-1979 (1992)
26. G. Triscone, J.-Y. Genoud, T. Graf, A. Junod, J. Muller, "Variation of the Superconducting Properties of $Bi_2Sr_2Ca_1Cu_2O_{8+x}$ with Oxygen Content," Physica C, **176**, 247-256 (1991)

INFLUENCE OF SILVER ON THE $(Bi,Pb)_2Sr_2Ca_2Cu_3O_x$ INCONGRUENT MELTING REACTION

Robert J. Moon, Kevin P. Trumble, and Keith J. Bowman
School of Materials Engineering, Purdue University
West Lafayette, IN 47907

ABSTRACT

The chemical and microstructural evolution of particulate lead doped $Bi_2Sr_2Ca_2Cu_3O_x$ ((Bi,Pb)2223) in contact with pure bulk silver at 830 to 905°C under flowing air and Ar-5%O_2 atmospheres has been studied. Specimens annealed under PO_2 and temperature conditions for which the (Bi,Pb)2223 phase incongruently melts, revealed that solid silver dissolved into the (Bi,Pb)2223 incongruent liquid, facilitating the incongruent melting reaction. Upon cooling, the dissolved silver precipitated out of the incongruent melt as equiaxed metallic silver particles and equiaxed copper rich particles.

INTRODUCTION

The powder-in-tube (PIT) method of producing silver-2223 composite wire and tape [1-5] are having limited success in maintaining the silver sheath uniformity due to intergrowths of the 2223 phase and secondary phase(s) into the silver sheath [6,7]. Larbalestier *et al.* [8] and Lelovic *et al.* [9] determined that the highest current carrying regions of these composites were located in a thin superconducting layer adjacent to the silver sheath, and because of this it is believed that a higher J_C can be obtained by a stronger 2223 phase alignment along a smooth silver surface. The non-uniformity of the silver sheath has been attributed to the mechanical deformation during tape processing but chemical interactions may have been overlooked.

The formation of the 2223 phase from the 2212 rich precursor powders used in PIT processing is believed to involve a liquid phase [10-13]. Several authors have shown that 2223 nucleates at and grows preferentially along the silver/2212 interface [7,14-16], and others have suggested that this results from silver lowering the melting temperature of the 2212 phase [17-19]. Silver has been shown to dissolve into the 2212 incongruent liquid [18,20] and non-uniform dissolution may cause the roughness of the interface. Additionally, the silver/2223 interface is atomically planar and free of secondary phases, except when subjected to too high of annealing temperature where the silver/2223 interface forms interfacial phases [21]. Unfortunately, there are few studies on silver-2223 interaction, so its effect on maintaining a smooth silver-2223 interface is not well understood.

Moon *et al.* [22,23] produced a (Bi,Pb)2223 PO_2-temperature stability diagram (Fig. 1) from an assessment of the literature and investigated the influence of silver on (Bi,Pb)2223 phase stability. Annealing within incongruent melting conditions silver dissolved into and diffused through the incongruent melt liquid, dictating the range of silver influence. This paper will provide further results demonstrating the interaction of silver in the (Bi,Pb)2223 incongruent melting reaction.

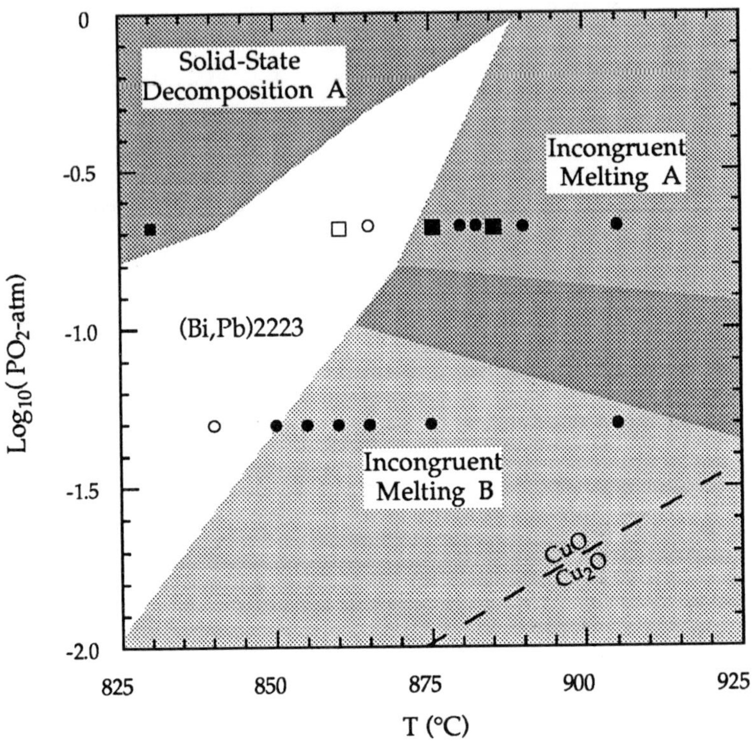

Figure 1. The PO_2-Temperature (Bi,Pb)2223 phase stability diagram compiled from a critical assessment of the literature. The extensions of the shaded regions to higher temperatures is shown only as a visual aide and have not been experimentally determined. The incongruent melting A reaction occurs by: (Bi,Pb)2223 ⇒ Liquid + $(Ca,Sr)_2CuO_3$ + $(Sr,Ca)CuO_2$ and the incongruent melting B reaction occurs by: (Bi,Pb)2223 ⇒ Liquid + $(Ca,Sr)_2CuO_3$ + O, where the liquid has a composition between 2201 and 2212. The filled markers represent specimen annealing conditions where (Bi,Pb)2223 decomposition was detected [22,23].

EXPERIMENTAL

All experiments were conducted using the same batch of commercial (Bi,Pb)2223 powder (mostly (Bi,Pb)2223 phase (see Fig. 3a), $Bi_{1.8}Pb_{0.33}Sr_{1.87}Ca_2Cu_3O_x$, platelet morphology ~2 to 20 μm X 0.5 μm, Seattle Specialty Ceramics, Seattle, WA). The (Bi,Pb)2223 powder was dry-pressed in a double-acting steel die at 140 MPa, producing 6 mm diameter x 1.5 mm thick pellets which were quartered axially using a razor blade.

Two forms of silver substrates were used in this study: foil (99.9%, 50 μm thick, Aldrich Chemical Company, Inc.) and rod (99.999%, 7 mm diameter, Johnson Matthey, Ward Hill, MA). Silver disks 1 mm thick were cut from the silver rod and one surface of each disk was mechanically polished through 1 μm diamond abrasive. The silver disks and foils were then annealed at 900°C in flowing air for 1 h to eliminate silver oxide or sulfide from the polished surface. High purity MgO substrates (99.9%, Baikowski) were used as control specimens due to minimal interaction of MgO with the BSCCO system [24].

One pellet quarter was placed at the center of a given substrate and the assemblages were placed onto an alumina support and inserted into a pre-heated alumina tube furnace at temperatures ranging from 750 to 905°C in either flowing air or Ar-5%O_2 atmospheres (1 atm total pressure) for 2 h. Figure 1 shows the annealing conditions superimposed on the PO_2-Temperature (Bi,Pb)2223 phase stability diagram. A unique feature of these experiments was that specimens had regions wherein the (Bi,Pb)2223 phase was either in contact with a given substrate (pellet bottom surface) or not in contact with the substrate (pellet top surface). This allowed for a direct comparison of differences in the (Bi,Pb)2223 phase stability due to the presence of the substrate.

Cylinders of commercial Ar-5%O_2 gas and dry air were used to maintain the two annealing atmospheres used in this study. Residual CO_2 adsorption onto the materials used in this study was not measured, but it is unlikely to have influenced the (Bi,Pb)2223 stability because the experiments were conducted in an open system [25-27]. A K-type thermocouple, periodically calibrated to a NIST-traceable standard, was placed within 1 cm of the specimen. The hot zone varied less than ± 2°C at 850°C over a 3 cm length. After annealing, specimens were cooled in ambient air. The pellet top and bottom pellet surfaces were examined by scanning electron microscopy (SEM) and energy dispersive spectroscopy (EDS). Specimens were also examined in cross-section by SEM, EDS, and electron probe microanalyzer (EPMA). The top 3/4 of pellets were crushed and XRD of this powder was used to study the phases present.

SPECIMEN CHARACTERIZATION

The SEM and EDS analyses were conducted using either a JEOL T-300 SEM operated at 20 kV with a Princeton-Gamma Tech EDS detector and IMIX imaging system or a JEOL JSM-35CF SEM at 20kV with JEOL EDS detector and Tracor Northern imaging system. X-ray powder diffraction was completed using a Siemens D-500 X-ray diffractometer with Cu Ka radiation.

Quantitative EPMA was conducted on a CAMECA SX 50 microprobe operated at 15 kV accelerating voltage, a 20 nA beam current, using the following standards (for elements): $AgBiSe_2$ (Ag, Bi), $SrTiO_3$ (Sr), $CaSiO_3$ (Ca), CuS (Cu, S), and PbS (Pb). A minimum of 5 point analyses were averaged for all phase compositions. Compound stoichiometries were determined on a basis of normalizing Cu = 1, 2, or 3, depending on the particular phase. The standard deviation for each normalized cation fraction was less than .08. Atomic percentages were obtained by assuming the undetected weight percentage was that of oxygen.

RESULTS AND DISCUSSION

Pellets annealed for 2 h on silver substrates exhibited more (Bi,Pb)2223 phase decomposition compared to pellets annealed on MgO substrates under identical conditions. SEM micrographs of the pellet top surface after annealing on silver for 2 h in air at 860°C, 875°C or 885°C are shown in Figure 2. Figure 3 shows the XRD of crushed pellets annealed under these conditions. Specimens annealed in Ar-5%O_2 atmospheres had similar (Bi,Pb)2223 phase stability and decomposition behavior to the specimens annealed in air with the lower PO_2 causing transitions to occur at lower temperatures (Fig. 1).

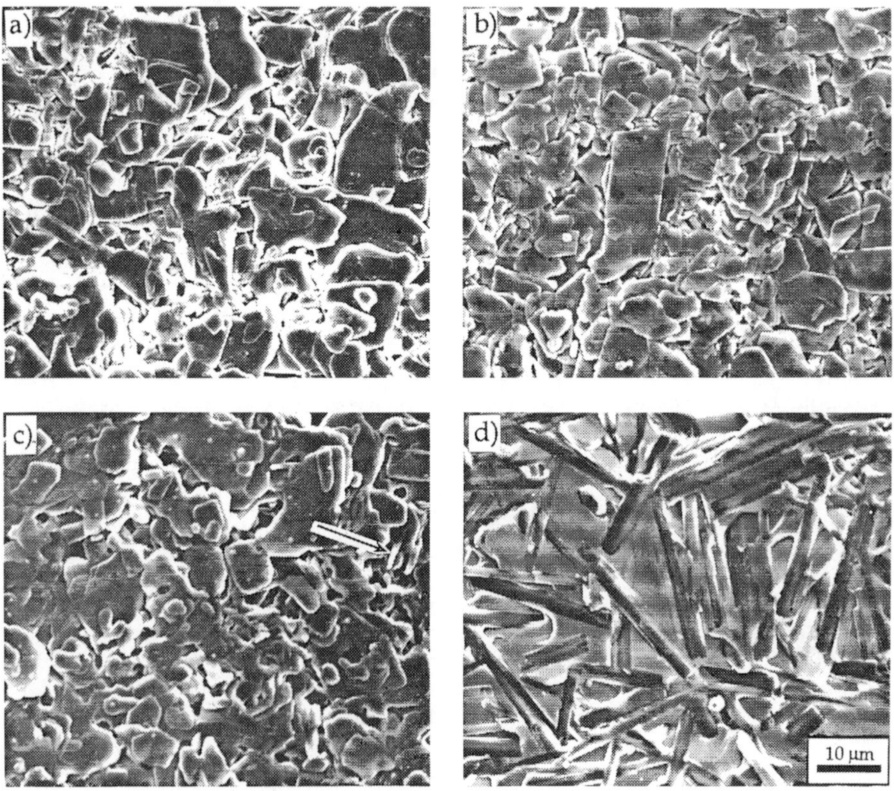

Figure 2. SEM micrographs showing top surfaces of (Bi,Pb)2223 pellets annealed on silver substrates for 2 h in air. a) Initial powder b) 860°C, c) 875°C, d) 885°C.

The 860°C specimen (stable annealing conditions) showed no changes in (Bi,Pb)2223 platelet morphology or phase composition. The 875°C specimen (incongruent melting conditions) demonstrated (Bi,Pb)2223 phase instability, where the top pellet surface showed platelet morphology changes with trace $(Ca,Sr)_2CuO_3$ needle formation and XRD detected the 2201 phase. The pellet bottom surface in contact with the silver substrate revealed a complete (Bi,Pb)2223 phase decomposition forming an extensive network of $(Ca,Sr)CuO_2$ needles, demonstrating an influence of silver on the (Bi,Pb)2223 decomposition.

For specimens annealed within the range of 885°C-905°C the entire pellet incongruently melted. X-ray diffraction detected $(Ca,Sr)_2CuO_3$, $(Ca,Sr)CuO_2$ and 2201 phases but did not detect the (Bi,Pb)2223 phase, confirming the complete (Bi,Pb)2223 decomposition. The $(Ca,Sr)_2CuO_3$ and $(Ca,Sr)CuO_2$ needle phases were surrounded by a liquid phase at annealing conditions, and on cooling the liquid solidified into a Ca-rich 2201 phase. The influence of silver on the bottom pellet surface could not be examined because the pellet bonded to the silver surface.

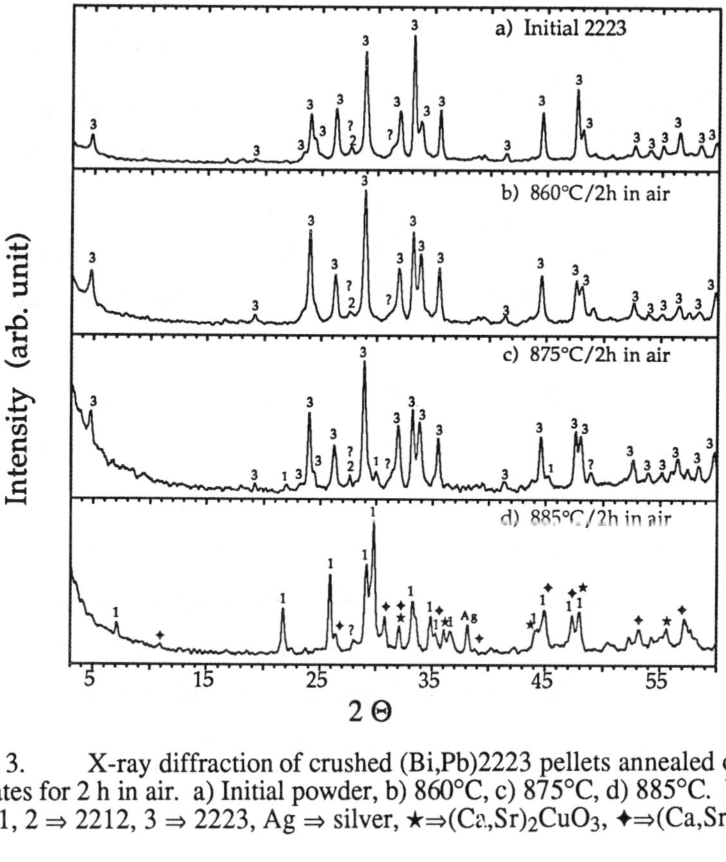

Figure 3. X-ray diffraction of crushed (Bi,Pb)2223 pellets annealed on silver substrates for 2 h in air. a) Initial powder, b) 860°C, c) 875°C, d) 885°C. Where 1 ⇒ 2201, 2 ⇒ 2212, 3 ⇒ 2223, Ag ⇒ silver, ★⇒$(Ca,Sr)_2CuO_3$, ✦⇒$(Ca,Sr)CuO_2$.

The influence of silver on the incongruent melting reaction was compared to (Bi,Pb)2223 pellets annealed on MgO substrates. SEM micrographs of the pellet top surface after annealing on MgO for 2 h in air at 875°C, 885°C or 890°C are shown in Figure 4. Figure 5 shows the XRD of crushed pellets annealed under these conditions. The top pellet surface and XRD of the 875°C specimen was similar to the pellets annealed on silver. However, the bottom pellet surface did not appear to have changed morphology, contrasting pellets annealed on silver. The 885°C specimen had a similar platelet morphology as the 875°C specimen and XRD detected a slight increase in the (Bi,Pb)2223 decomposition but this was much less than the decomposition that occurred to pellets annealed on silver substrates. Specimens annealed at 890°C exhibited an increase in the (Bi,Pb)2223 decomposition, where the XRD detected 2223, 2212, and 2201 phases, but the extent of decomposition was much less than pellets annealed on silver.

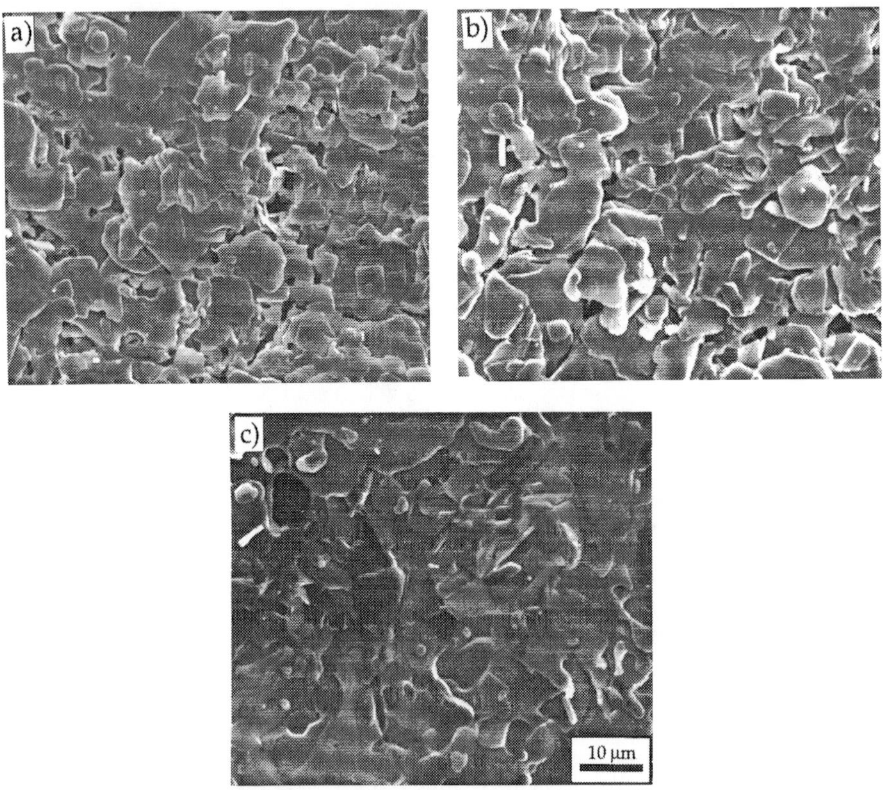

Figure 4. SEM micrographs showing top surfaces of (Bi,Pb)2223 pellets annealed on MgO substrates for 2 h in air. a) 875°C, b) 885°C, c) 890°C.

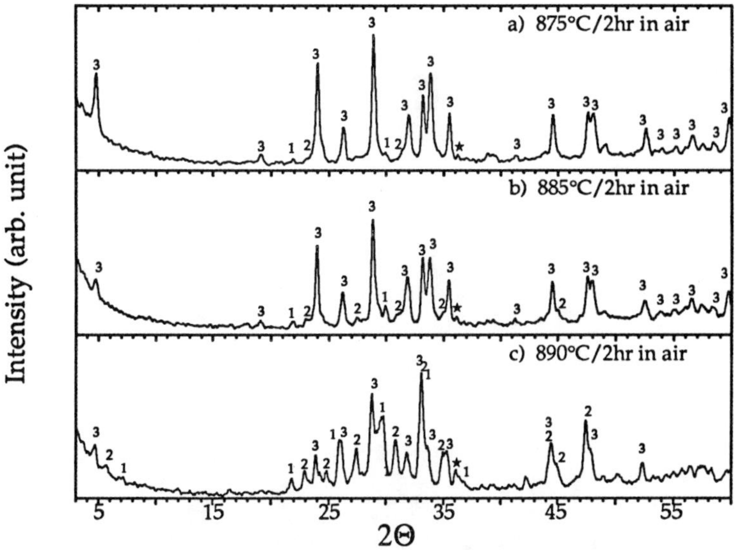

Figure 5. X-ray diffraction of (Bi,Pb)2223 pellets annealed on MgO. Note that the amount of (Bi,Pb)2223 decomposition at 885 and 890°C are much less than pellet annealed on silver at identical conditions. a) 875°C/2h in air, b) 885°C/2h in air, c) 890°C/2h in air.

The influence of silver on the incongruent melting reaction was demonstrated by annealing (Bi,Pb)2223 pellets on a MgO disk and a silver disk, side-by-side at 890°C/2h in air. The pellet on silver exhibited complete incongruent melting (Fig. 3d), whereas the pellet on MgO showed significantly decreased decomposition (Fig. 5b).

The increased amount of (Bi,Pb)2223 decomposition for pellets annealed on silver substrates was caused by silver dissolving into the pellets. Cross-sections of pellets that were incongruently melted on silver substrates revealed a uniform dispersion of equiaxed silver particles (<10μm) throughout the entire pellet. Silver was observed in all pellets that were annealed for 2 h within 885 to 905°C in air or 865 to 905°C in Ar-5%O_2. X-ray diffraction of powder scraped from the top 3/4 of the pellet detected the (111) peak of metallic silver, which is non-overlapping with any of the BSCCO phases (Fig. 3d).

A cross-section of an incongruently melted pellet that was annealed on a silver foil at 905°C/2h in air is shown in Figure 6, where the specimen configuration isolated the incongruent melting liquid from the pellet by placing the pellet over a ~0.5 mm diameter hole in the silver foil. The liquid phase flowed through the hole, collecting on the other side of the foil and thereby isolating it from the pellet. The liquid phase also spread across the entire foil top and bottom

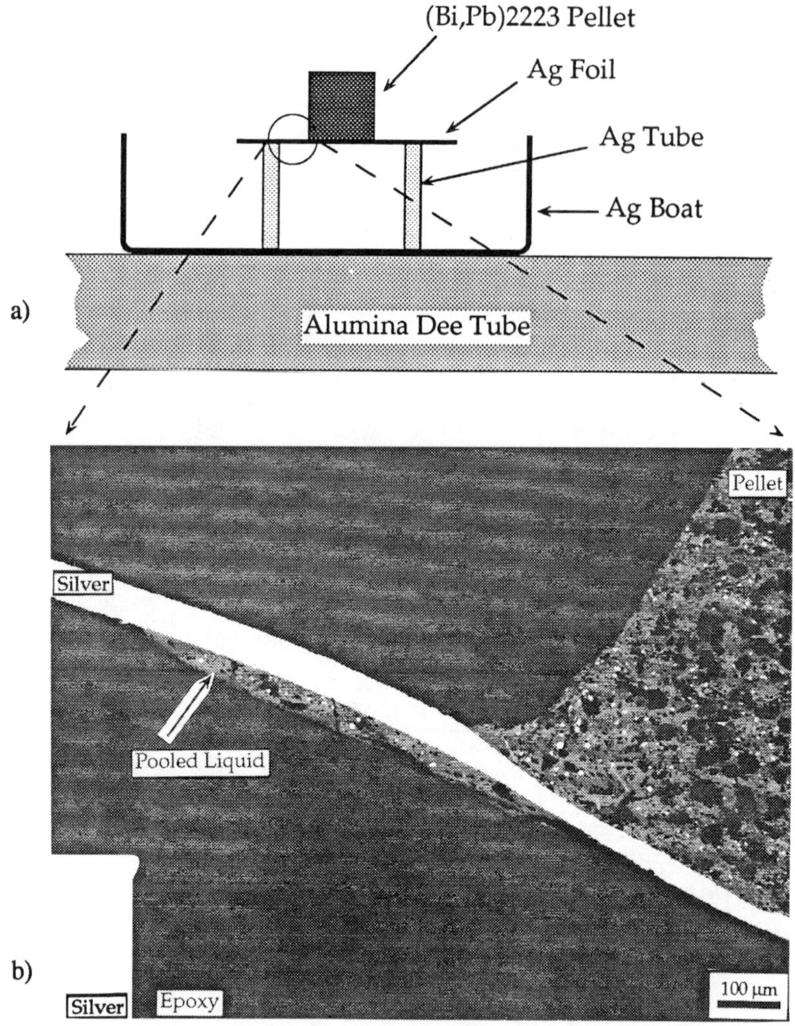

Figure 6 a) Schematic of specimen configuration to isolate the incongruent melt liquid. b) Optical micrographs of the silver foil left side after annealing at 905°C/2h in air showing the isolated liquid from the pellet. Equiaxed silver particles precipitated from the incongruent melt liquid on cooling.

surface, having an apparent contact angle with the silver foil of <<90°. The silver precipitates (white) in the pellet and the isolated incongruent melt liquid demonstrate that silver dissolved into the incongruent melt. The reduction in the silver foil thickness is also apparent.

Enlargements of this liquid/silver foil region are shown in Figure 7, where the solidified incongruent melt liquid exhibited compositional variations (confirmed by EPMA). The large (>20 μm), faceted platelets having a composition of $(Bi,Pb)_{2.5}Sr_{2.3}Ca_{0.8}Cu_2O_x$ precipitated first, producing a second liquid phase. As the remaining liquid solidified ($(Bi,Pb)_{3.8}Sr_2Ca_{1.3}Cu_2O_x$), equiaxed metallic silver, equiaxed Cu-Ag oxide (~7 at% Ag), and $(Ca,Sr)CuO_2$ phases precipitated in an apparent eutectic type of solidification. These precipitates within the solidified liquid phase were also observed throughout the incongruently melted pellet, but it is unclear why XRD detected the 2201 phase as a major phase and not one of the majority solidified incongruent melt phases (Fig. 3d).

An estimate of the amount of silver dissolved into the incongruent liquid was completed by averaging 20 non-overlapping EMPA area scans of the isolated solidified liquid shown in Figure 7. The amount of silver that dissolved into the incongruent melt liquid when annealing at 905°C/2h in air was ~ 4.5 at%.

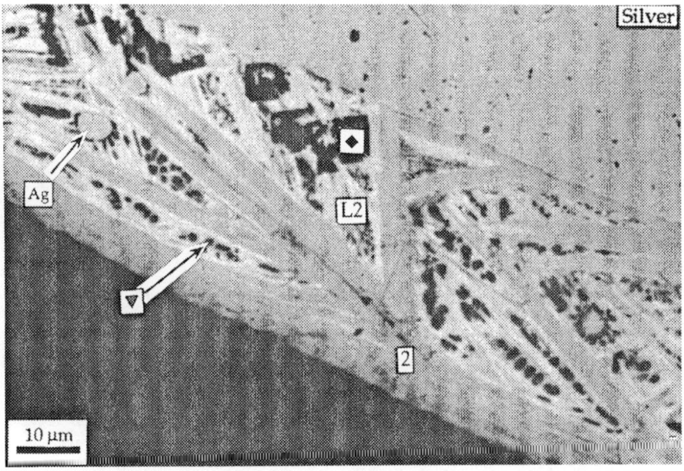

Figure 7. Backscattered image of a pellet cross-section that was annealed at 905°C/2h in air showing composition variation within the solidified incongruent melt liquid isolated from the pellet. Where Ag⇒ silver, 2⇒ $(Bi,Pb)_{2.5}Sr_{2.3}Ca_{0.8}Cu_2O_x$, ✦⇒ $(Ca,Sr)CuO_2$, L2 ⇒ $(Bi,Pb)_{3.8}Sr_2Ca_{1.3}Cu_2O_x$, ▼⇒ (Cu, 7 at% Ag) oxide.

Specimens annealed at 890°C in air for 15, 30, 60 or 120 min demonstrated a time dependent range of silver influence on the pellet incongruent melting decomposition. Initially incongruent melting only occurred at the pellet bottom surface in contact with silver, as observed in the 15 min specimen, but after annealing for 30 min pellet incongruent melting moved into the interior of the pellet and after 60 and 120 min of annealing the entire pellet incongruently melted. For specimens annealed at 905°C in air for either 15, 30, or 120 min, the entire pellet incongruently melted where the pellet cross-section showed a uniform distribution of equiaxed metallic silver particles throughout the pellet. However, the silver particle size of the 15 min specimen was less than half the size of the 120 min specimen and the silver foil thinning underneath the pellet was not as great, demonstrating continued silver dissolution with time.

SUMMARY

The observed onset of (Bi,Pb)2223 incongruent melting was in good agreement with the stability diagram produced from a critical assessment of the literature. The (Bi,Pb)2223 incongruent melting behavior was strongly influenced by the silver substrate, where it was demonstrated that silver dissolves into the incongruent melting liquid. Pellets annealed on MgO substrates at identical conditions had significantly less amount of (Bi,Pb)2223 decomposition within the 2 h anneals, demonstrating that silver facilitates the (Bi,Pb)2223 decomposition reaction. The range of silver influence on (Bi,Pb)2223 decomposition and the amount of silver dissolved into the incongruent melt liquid was dependent on the annealing temperature and time. On cooling the incongruent melt liquid solidified and the dissolved silver precipitated out as equiaxed metallic silver and equiaxed copper rich particles.

REFERENCES

[1]J. P. Singh, J. Joo, N. Vasanthamohan, and R. B. Poeppel, "Role of Ag Additions in the Microstructural Development, Strain Tolerance, and Critical Current Density of Ag-Sheathed BSCCO Superconducting Tapes," *J. Mater. Res.* **8** [10], 2458-2464 (1993).

[2]J. Joo, J. P. Singh and R. B. Poeppel, "Effects of Thermomechanical Treatment on Phase Development and Properties of Ag-Sheathed Bi(Pb)-Sr-Ca-Cu-O Superconducting Tapes," *Supercond. Sci. Technol.* **5**, 421-428 (1993).

[3]J. M. Yoo and K. Mukherjee, "X-ray Polefigure Analyses of Texturing and Textural Hardening Observed in Mechanical Deformation of $Bi_{1.6}Pb_{0.4}Sr_2Ca_2Cu_3O_z$ Superconductor HIP Cladded on Ag Substrate," *Physica C* **222**, 241-251 (1994).

[4]S. Ahzi, R. J. Asaro, and D. M. Parks, "Application of Crystal Plasticity Theory for Mechanically Processed BSCCO Superconductors," *Mech. Mats.* **15**, 201-222 (1993).

[5]P. Halder, and L. Motowidlo, "Processing High Critical Current Density Bi-2223 Wires and Tapes," *JOM* **44** (10), 54-58 (1992).

[6]A. Umezawa, Y. Feng, H. S. Edelman, Y. E. High, D. C. Larbalestier, Y. S. Sung and E. E. Hellstrom, "Electromagnetic Granularity, Critical Current Density, and Low- Tc Phase Formation at the Grain Boundaries in (Bi, $Pb)_2Sr_2Ca_2Cu_3O_x$ Silver-Sheathed Tapes," *Physica C* **198**, 261-272 (1992).

[7]Y. Feng, Y. E. High, D. C. Larbalestier, Y. S. Sung and E. E. Hellstrom, "Evidence for Preferential Formation of the $(Bi, Pb)_2Sr_2Ca_2Cu_3O_x$ Phase at the Ag Interface in Ag-Sheathed $(Bi, Pb)_2Sr_2Ca_2Cu_3O_x$ Tapes," *Appl. Phys. Lett.* **62** [13], 1553-1555 (1993).

[8]D. C. Larbalestier, X. Y. Cai, Y. Feng, H. Edelman, A. Umezawa, G. N. Riley Jr., and W. L. Carter, "Position-Sensitive Measurements of the Local Critical Current Density in Ag Sheathed High-Temperature Superconductor $(Bi,Pb)_2Sr_2Ca_2Cu_3O_y$ Tapes," *Physica C* **221**, 299-303 (1994).

[9]M. Lelovic, P. Krishnaraj, N. G. Eror, and U. Balachandran, "Minimum Critical Current Density of 10^5 A/cm^2 at 77K in the Thin Layer of $Bi_{1.8}Pb_{0.4}Sr_{2.0}Ca_{2.2}Cu_{3.0}O_y$ Superconductor Near the Ag in Ag-Sheathed Tapes," *Physica C* **242**, 246-250 (1995).

[10]P. E. D. Morgan, R. M. Housley, J. R. Porter and J. J. Ratto, "Low Level Mobile Liquid Droplet Mechanism Allowing Development of Large Platelets of High-T_c 'Bi-2223' Phase Within a Ceramic," *Physica C* **176**, 279-284 (1991).

[11]P. E. D. Morgan, J. D. Piche and R. M. Housley, "Use of a Thermal Gradient to Study the Role of Liquid Phase During the Formation of Bismuth High Temperature Superconductors," *Physica C* **191**, 179-184 (1992).

[12]Y. L. Chen and R. Stevens, "2223 Phase Formation in Bi(Pb)-Sr-Ca-Cu-O: II, The Role of Temperature-Reaction Mechanism," *J. Am. Ceram. Soc.*, **75** [5], 1150-1159 (1992).

[13]Yamada, Y., B. Obst and R. Flukiger, "Microstructural Study of Bi (2223)/Ag Tapes with J_c(77K, 0T) Values of up to $3.3X10^4$ A cm^{-2}," *Supercond. Sci. Technol.* **4**, 165-171 (1991).

[14]Y. E. High, Y. Feng, Y. S. Sung, E. E. Hellstrom, and D. C. Larbalestier, "A Study of the Competitive Reactions Occurring During the Heat Treatment of Silver-Sheathed $Bi_{1.8}Pb_{0.4}Sr_{2.0}Ca_{2.2}Cu_{3.0}O_y$ Aerosol Spray Pyrolysis Powder," *Physica C* **220**, 81-92 (1994).

[15]J. S. Luo, N. Merchant, V. A. Maroni, G. N. Riley, Jr. and W. L. Carter, "Influence of Silver Cladding on the Formation and Alignment of the $(Bi_{2-x}Pb_x)Sr_2Ca_2Cu_3O_{10+d}$ Phase," *Appl. Phys. Lett.* **63** [5], 690-692 (1993).

[16]M. Lelental, T. N. Blanton, C. L. Barnes and H. J. Romanofsky, "Semiquantitative Assessment of 2223/2212 Phase Homogeneity in Bi-(Pb)-Sr-Ca-Cu-O Superconducting Thin Films Using XRD Techniques," *Physica C* **193**, 395-400 (1992).

[17]Y. C. Guo, H. K. Liu and S. X. Dou, "Phase Evolution in Silver-Doped BiPbSrCaCuO(2223)/Ag Superconducting Composites," *J. Mater. Res.* **8** [9], 2187-2190 (1993).

[18]J. L. MacManus-Driscoll, J. C. Bravman, R. J. Savoy, G. Gorman, and R. B. Beyers, "Effects of Silver and Lead on the Phase Stability of $Bi_2Sr_2Ca_1Cu_2O_{8+x}$ and $Bi_2Sr_2Ca_2Cu_3O_{10+x}$ above and below the Solidus Temperature," J. Am. Ceram. Soc. **77** [9] 2305-2313 (1994).

[19]Sung, Y. S. and E. E. Hellstrom, "A Role of Ag in the Reaction Formation forming $(Bi,Pb)_2Sr_2Ca_2Cu_3O_x$, $(Bi,Pb)_2Sr_2CaCu_2O_x$, $(Sr,Ca)_2CuO_3$ and CuO," *Physica C* **255**, 266-274 (1995).

[20]Ray II, R. D. and E. E. Hellstrom, "Ag-Clad Bi-Sr-Ca-Cu-O Wires: I. Phase Study of Fully Processed 2: 2: 1: 2 wires," *Physica C* **172**, 435-440 (1991).

[21]O. Eibl, "The High-Tc Compound $(Bi,Pb)_2Sr_2Ca_2Cu_3O_{10+d}$: Features of the Structure and Microstructure Relevant for Devices in Magnet and Energy Technology," *Supercond. Sci. Technol.* **8**, 833-861 (1995).

[22] R. Moon, K. Trumble and K. Bowman, "Thermal Instability of $(Bi,Pb)_2Sr_2Ca_2Cu_3O_x$ in Contact with Silver," Submitted to *J. Am. Ceram. Soc.*.

[23]R. Moon, "Thermal Instability of $(Bi,Pb)_2Sr_2Ca_2Cu_3O_x$ in Contact with Silver," M.S. Thesis, Purdue University, Dec. 1996.

[24]S. Dou, H. Liu, S. Guo, K. Easterling and J. Milael, "Superconductivity in the Bi-Pb-Sr-Ca-Cu-O System Wtih Oxide Additions," *Supercond. Sci. Technol.* **2**, 274-278 (1989).

[25]R. H. Arendt, M. F. Gargauskas, E. L. Hall, K. W. Lay, and J. E. Tkaczyk, "The Role of Ag in the Preparation of Metal Sheathed $(Bi,Pb)_2Ca_2Sr_2Cu_3O_z$ Tapes," *Physica C* **194**, 393-396 (1992).

[26]W. Zhang and E. Hellstrom, "The Influence of Carbon on Melt Processing Ag Sheathed $Bi_2Sr_2CaCu_2O_8$ Tape," *Physica C* **234**, 137-145 (1994).

[27]L. Margulies, K. Dennis, R. Hofer, M. Kramer, R. McCallum, "The Effect of Carbon on the Peritectic Melting of Bi2212 and Bi2212+Ag," *Physica C* **264**, 133-136 (1996).

INFLUENCE OF FIRING PROCEDURES IN THE SOL-GEL SYNTHESIS OF $YBa_2Cu_3O_{6+x}$ SUPERCONDUCTORS

A. Manthiram, C. Tsang and C. Ramanujan
Center for Materials Science and Engineering, ETC 9.104
The University of Texas at Austin
Austin, TX 78712

ABSTRACT

The influence of firing procedures on the crystal chemistry and superconducting properties of $YBa_2Cu_3O_{6+x}$ has been investigated systematically by firing at different temperatures a citrate derived precursor. While a tetragonal nonsuperconducting phase with oxygen contents of about 6.6 is formed for firing temperatures $750 \leq T \leq 800$ °C, orthorhombic superconducting phases are formed for $T > 830$ °C. The superconducting diamagnetic signal of the orthorhombic phases increases with increasing firing temperatures. More importantly, the tetragonal phase formed at the intermediate firing temperature (800 °C) is found to have significant influence on the structure and properties of the orthorhombic superconducting phases formed by subsequently firing at $T > 900$ °C. It leads to oxygen disordering, lower oxygen contents and inferior superconducting properties in the orthorhombic phases.

INTRODUCTION

$YBa_2Cu_3O_{6+x}$ has an intergrowth structure (Fig. 1) in which the superconductively active layers consisting of two $Cu(2)O_2$ sheets and the inactive layers consisting of a single $Cu(1)O_x$ chain planes and two BaO rocksalt planes alternate along the c-axis. The intergrowth sequence along the c-axis is

$$|\,Cu(2)O_2\text{-}Y\text{-}Cu(2)O_2\,|\,BaO\text{-}Cu(1)O_x\text{-}BaO\,| \qquad (1)$$

The oxygen content in the $Cu(1)O_x$ chain planes is variable for $0 \leq x \leq 0.96$ depending upon the synthesis and processing conditions [1, 2]. The sample is superconducting for $0.4 < x \leq 0.96$ and semiconducting for $0 \leq x < 0.3$ [3, 4]. In addition, the oxygen ordering in the $Cu(1)O_x$ planes play a profound role on the superconducting properties. While a perfect ordering of oxygen atoms along the b-

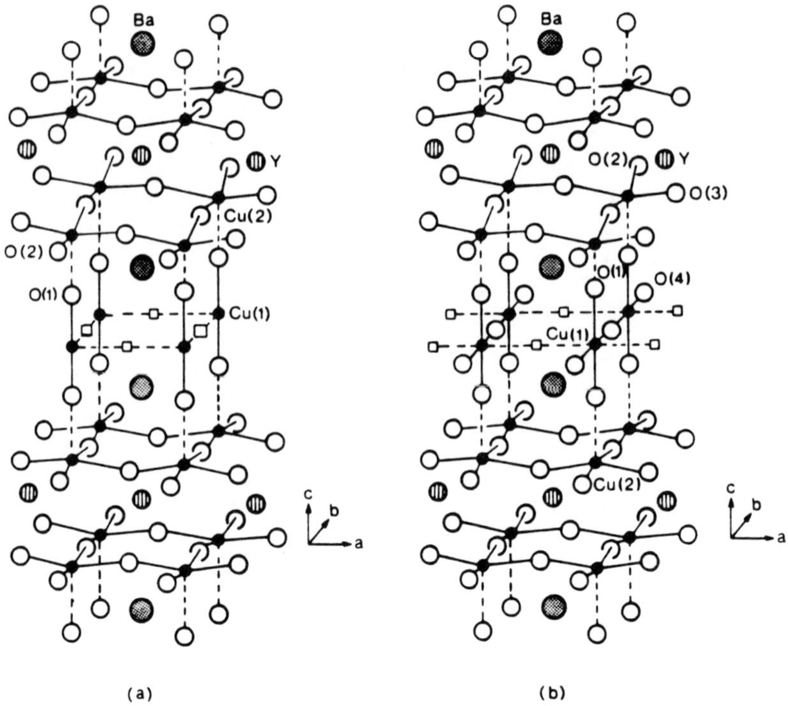

Fig. 1 Crystal structures of (a) tetragonal $YBa_2Cu_3O_6$, and (b) ideal, orthorhombic $YBa_2Cu_3O_7$.

axis leads to orthorhombicity and good superconducting properties with a maximum $T_c = 92$ K for $x \approx 0.96$, any disordering of oxygen atoms in the $Cu(1)O_x$ planes results in a decrease in the degree of orthorhombicity and mobile hole concentration in the $Cu(2)O_2$ sheets, which in turn results in a degradation or vanishing of superconducting properties [5-7].

$YBa_2Cu_3O_{6+x}$ is generally synthesized by reacting required quantities of the constituent oxides or carbonates at elevated temperatures $T \approx 950$ °C with intermittent grindings. Such high temperature procedures generally lead to larger grain size. Alternatively, use of precursors in which the three constituent metal ions -- Y^{3+}, Ba^{2+}, and Cu^{2+} -- are intimately mixed not only can bring down the firing temperatures but can also offer smaller grain size, higher driving force for densification and better homogeneity in the final product. However, the synthesis of $YBa_2Cu_3O_{6+x}$ at lower temperatures $T \leq 800$ °C has been found to give a tetragonal, nonsuperconducting phase with a significantly lower oxygen contents $(6+x) = 6.7$ [7, 8].

Although it is well recognized now that a firing temperature T > 900 °C is essential in order to achieve good superconducting properties, the consequences that may be introduced by a prefiring of the intimately mixed precursors at moderate temperatures 700 < T < 900 °C before finally firing at T > 900 °C have rarely been paid attention. An understanding of the influence of such an intermediate firing step will be valuable in optimizing the processing procedures involved in solution-based synthesis methods to achieve better properties in the superconducting phase. Specifically, it is critical that one needs to be aware of the advantages and disadvantages of and the factors that may play a role in the solution-based, low temperature synthesis of $YBa_2Cu_3O_{6+x}$. This paper presents (i) the synthesis of $YBa_2Cu_3O_{6+x}$ using a citrate-based sol-gel precursor, (ii) the variation of crystal chemistry and oxygen content with synthesis temperature, and (iii) the influence of an intermediate firing step at 800 °C on the crystal chemistry, oxygen content and superconducting properties of the superconducting phase obtained by subsequently firing at T > 900 °C.

EXPERIMENTAL

A citrate-based sol-gel method [9] was used to synthesize the $YBa_2Cu_3O_{6+x}$ samples. To an aqueous solution consisting of $Y(CH_3COO)_3.4H_2O$, $Ba(CH_3COO)_2$ and $Cu(CH_3COO)_2.2H_2O$ in the molar ratio 1:2:3, citric acid (1.9 g per g of $Ba(CH_3COO)_2$) and ethylene glycol (10 ml per mmol of $Y(CH_3COO)_3.4H_2O$) were added under constant stirring. The resulting solution was then heated on a hot plate until a brown black solid mass was obtained. The citrate precursor thus obtained was then decomposed at about 500 °C in air to remove most of the organics. The black powder so obtained was then pelletized and fired at various temperatures ranging from 750-950 °C for 1-3 days to obtain $YBa_2Cu_3O_{6+x}$.

Crystal chemical analysis and phase identifications were carried out by X-ray powder diffraction. Oxygen contents were determined by iodometric titration as reported earlier [1]. Thermogravimetric analysis (TGA) plots to evaluate the oxygen loss behavior were recorded with a Perkin-Elmer Series 7 Thermal Analysis system at a heating rate of 1 °C/min in N_2 atm. Superconducting properties were evaluated with a SQUID magnetometer with a field of 10 Oe.

RESULTS AND DISCUSSION

The citrate-based precursor was fired at various temperatures in order to understand the influence of firing temperatures on the structure and properties. X-ray powder diffraction of the products reveals that the reaction is incomplete at temperatures T < 750 °C and a firing temperature T ≥ 750 °C is necessary to obtain the $YBa_2Cu_3O_{6+x}$ phases. Fig. 2 gives the X-ray diffraction patterns for various firing temperatures and Fig. 3 shows the variation of lattice parameters with synthesis temperature. The data shows that the $YBa_2Cu_3O_{6+x}$ samples are

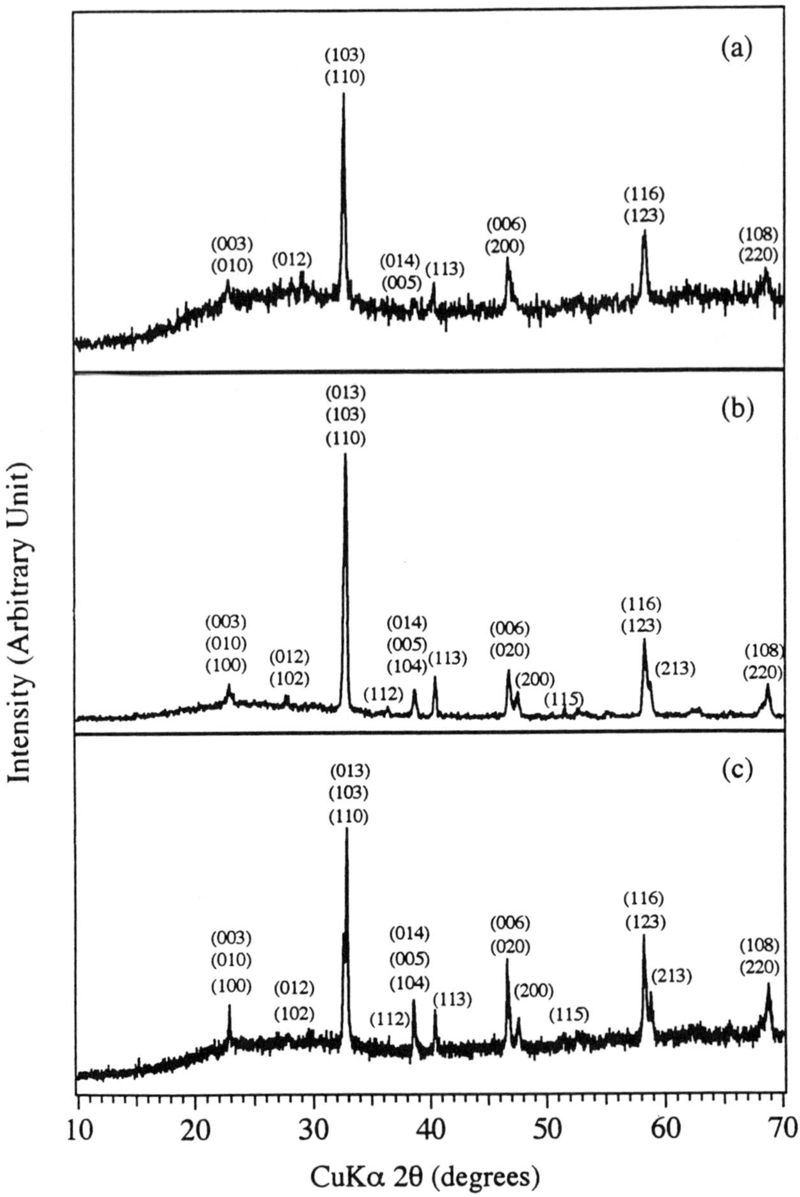

Fig. 2 X-ray diffraction patterns of $YBa_2Cu_3O_{6+x}$ samples obtained by firing the citrate precursor at different temperatures: (a) 790 °C, (b) 830 °C, and (c) 930 °C.

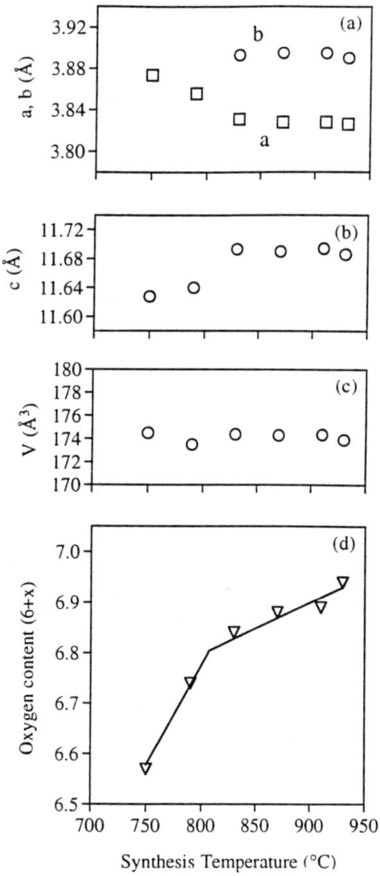

Fig. 3 Variations of lattice parameters and oxygen content of $YBa_2Cu_3O_{6+x}$ with synthesis temperature: (a) a and b parameters, (b) c parameter, (c) cell volume, and (d) oxygen content.

tetragonal for firing temperatures T ≤ 800 °C, indicating that the oxygen atoms are disordered in the $Cu(1)O_x$ planes. As the synthesis temperature increases above 800 °C, the samples become orthorhombic as indicated by a clear splitting between the (020) and (200) reflections in Fig. 2, indicating a suppression of oxygen disorder. The orthorhombic phases have a slightly larger c parameter than the tetragonal phases (see later).

Fig. 3 also shows the variation of oxygen content with synthesis temperature. The tetragonal samples have lower oxygen contents than the orthorhombic samples. The oxygen content increases with increasing synthesis temperature, but the rate of increase is slower for the orthorhombic samples compared to that for the tetragonal samples. Neutron diffraction experiments carried out a few years ago on the tetragonal samples suggested some oxygen vacancies in the BaO planes [10]. However, the diffraction peaks were broad and, therefore, precise refinement of the neutron data could not be obtained. The presence of oxygen vacancies in the BaO planes may be the origin of a lower oxygen content and smaller c-parameter in the tetragonal phases. Alternatively, the presence of some carbonate groups [11, 12] in the tetragonal phases may also influence the oxygen content values and lower the c parameter. Also, the presence of some unreacted impurity phases such as $BaCO_3$ for lower firing temperatures may influence the oxygen content values.

TGA plots recorded in N_2 atmosphere shows that the tetragonal samples are remarkably stable up to about 650 °C without much weight loss. On the other hand, the orthorhombic samples prepared at $T \geq 900$ °C are found to lose oxygen above about 300 °C with a sharp weight loss in the range 350-600 °C similar to that found with the orthorhombic samples obtained by conventional ceramic procedures [1]. These observations suggest that the disordered oxygen atoms in the $Cu(1)O_x$ planes are less mobile and more difficult to remove than the ordered oxygen atoms in the orthorhombic samples. Similar observations have also been found in the system $YBa_{2-y}La_yCu_3O_{6+x}$ [13]. The degree of orthorhombicity decreases with increasing y and the samples become tetragonal for $y > 0.3$ in this system and, therefore, the oxygen atoms become less mobile with increasing value of y [5].

Magnetic measurements show that the tetragonal phases are nonsuperconducting, while the orthorhombic phases are superconducting. These observations are consistent with earlier resistivity results obtained with samples prepared by a decomposition of coprecipitated oxalates [7]. A disordering of the oxygen atoms in the $Cu(1)O_x$ planes results in a trapping of the mobile holes from the superconductively active $Cu(2)O_2$ sheets and a suppression of superconducting properties. The orthorhombic sample synthesized at 830 °C has a diamagnetic onset temperature of 82 K while all other orthorhombic samples have an onset temperature of about 90 K. Although the onset temperatures of the orthorhombic samples do not vary significantly with the synthesis temperature, the Meissner fraction measured at 5 K increases with increasing synthesis temperature as shown in Fig. 4. An increase in the oxygen content and the formal oxidation state of Cu as well as an increasing degree of oxygen ordering with increasing synthesis temperature lead to better superconducting properties. *i. e.* an improvement in the ordering of oxygen atoms results in a less trapping of the mobile holes from the $Cu(2)O_2$ sheets.

The next question that needs to be addressed is whether the tetragonal nonsuperconducting phase formed at an intermediate temperature $T \leq 800$ °C has any possible influence on the structure and superconducting properties of the orthorhombic superconducting phase formed by subsequent firing at higher temperatures $T > 800$ °C. To establish this point, the following set of experiments

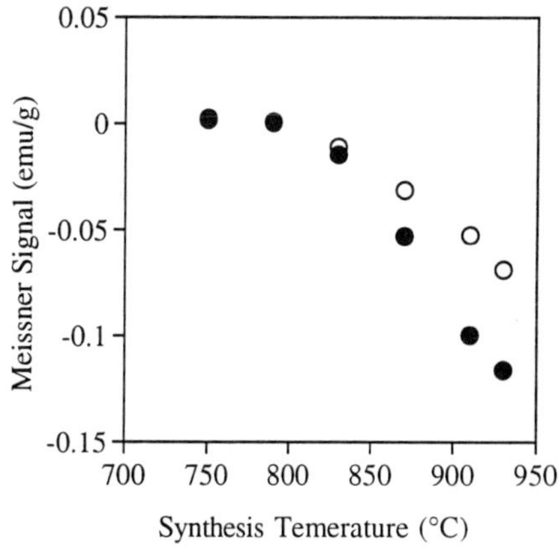

Fig. 4 Variation of Meissner signal measured at 5 K with the synthesis temperature of $YBa_2Cu_3O_{6+x}$. The closed and open symbols refer, respectively, to zero-field cooled and field-cooled data.

were carried out. The precursor obtained after firing at 500 °C was first fired at 800 °C for about 3 days to obtain the tetragonal nonsuperconducting phase. This tetragonal phase so obtained was then fired at 920 and 950 °C for 24 hr to give the orthorhombic superconducting phases, which we designate, respectively, as samples A and B. The precursor obtained at 500 °C was also fired straight away at 920 and 950 °C for 24 hr without the intermediate firing at 800 °C to give the orthorhombic superconducting phases, which we designate, respectively, as samples C and D. To avoid any variations that may occur in experimental conditions, samples A and C were, in fact, fired together at 920 °C and similarly samples B and D were fired together at 950 °C. The four orthorhombic samples A, B, C and D were then annealed in oxygen at 450 °C for 12 hr, held at 350 °C for 6 hr, and cooled to room temperature. The crystal chemistry and superconducting properties of the oxygen annealed samples were then characterized and compared.

Table I compares the lattice parameters and oxygen content values of all the four samples. Samples C and D have a higher degree of orthorhombicity than samples A and B. This indicates that samples C and D have a better oxygen ordering in the $Cu(1)O_x$ planes than samples A and B. Also, the observed lattice parameters and degree of orthorhombicity of samples C and D are in good agreement with those found [1] for the samples synthesized previously by conventional ceramic procedures of firing the constituent oxides and carbonates at 920-950 °C. This suggests that the samples C and D have nearly a perfect ordering of oxygen atoms along the b axis of the $Cu(1)O_x$ planes and samples A and B have still some disordering of oxygen atoms. In addition, samples A and B have lower c parameters and significantly lower oxygen contents than samples C and D. The

Table I. Comparison of lattice parameters and oxygen contents of $YBa_2Cu_3O_{6+x}$ samples prepared with and without an intermediate firing step at 800 °C

Sample	Lattice parameters (Å)			orthorhombicity	Oxygen
	a	*b*	*c*	*b/a*	content (6+x)
A	3.835(5)	3.883(5)	11.664(6)	1.013	6.61
B	3.828(2)	3.879(2)	11.648(3)	1.013	6.72
C	3.821(4)	3.884(4)	11.686(6)	1.017	6.89
D	3.826(1)	3.889(1)	11.686(1)	1.017	6.91

lower c parameter and oxygen content suggest that samples A and B may still have some oxygen vacancies in the BaO planes or carbonate groups in the lattice; *i. e.* the oxygen vacancies in the BaO planes or carbonate groups in the tetragonal sample prepared at 800 °C are not completely eliminated even after firing at 920-950 °C.

The observation of a lower c parameter for lower oxygen content is in contrast to an increase in the c-parameter found upon extracting oxygen from the orthorhombic $YBa_2Cu_3O_{6.94}$ that was obtained by conventional ceramic procedure [1]. This difference can be understood by considering the differences involved in the oxygen insertion/extraction processes. In the conventional ceramic method, the orthorhombic phase is formed first with no oxygen vacancies in the BaO planes or carbonate groups. Oxygen extraction from this material then involves creation of vacancies in the $Cu(1)O_x$ planes and a reduction of Cu^{2+} to Cu^+. An increase in the size of Cu^+ is reflected in an increase in the overall c parameter with decreasing oxygen content. In the present investigation, a tetragonal $YBa_2Cu_3O_{6.6}$ is formed first with oxygen vacancies in the BaO planes or carbonate groups in the lattice and a disordering of oxygen in the $Cu(1)O_x$ planes. The presence of oxygen vacancies in the BaO planes or carbonate groups leads to a lower c parameter in $YBa_2Cu_3O_{6.6}$. Subsequent firing of the tetragonal $YBa_2Cu_3O_{6.6}$ at higher temperature $T > 800$ °C involves a filling up of oxygen vacancies in the BaO planes or elimination of carbonate groups, which leads to an in increase in the c parameter with increasing oxygen content despite an increase in the overall oxidation state of Cu. But the oxygen vacancies in the BaO planes may not yet be completely filled or the carbonate groups may not yet be completely eliminated. The firing at higher temperature of the tetragonal $YBa_2Cu_3O_{6.6}$ also involves a removal of oxygen from the disordered $Cu(1)O_x$ planes, followed by insertion of oxygen atoms predominantly along the b axis during the subsequent cooling process. But the presence of still some oxygen vacancies in the BaO planes or carbonate groups may suppress the oxygen ordering in the $Cu(1)O_x$ planes during the cooling process.

Fig. 5 compares the superconducting properties of samples A and C. Sample C shows much sharper superconducting transition and larger diamagnetic signal than sample A. Similarly sample D was found to show sharper transition and larger signal than sample B. These observations are in accordance with the differences in crystal chemistry and oxygen contents. A slight disordering of the oxygen atoms and a lower oxygen content (or oxidation state of Cu) result in broader transitions and lower diamagnetic signals in samples A and B. Also, inhomegeneities in oxygen distribution in samples A and B may lead to broader transitions. These results clearly indicate that a prolonged intermediate firing at around 800 ^{o}C before firing subsequently at 920-950 ^{o}C can degrade the superconducting properties.

CONCLUSIONS

A systematic investigation of the crystal chemistry and superconducting properties of $YBa_2Cu_3O_{6+x}$ as a function of firing temperatures has led to several important conclusions. A firing temperature T > 900 ^{o}C is essential to achieve oxygen ordering in the chains and good superconducting properties. Firing temperatures T < 800 ^{o}C leads to oxygen disorder and nonsuperconducting properties. Although lower firing temperatures may be preferred to avoid grain growth, such a strategy in the case of $YBa_2Cu_3O_{6+x}$ leads to unwanted consequences. Prefiring of the precursors at intermediate temperatures 700 < T < 900 ^{o}C before finally firing at T ≈ 950 ^{o}C should be avoided since such a process leads to oxygen disorder and inferior superconducting properties.

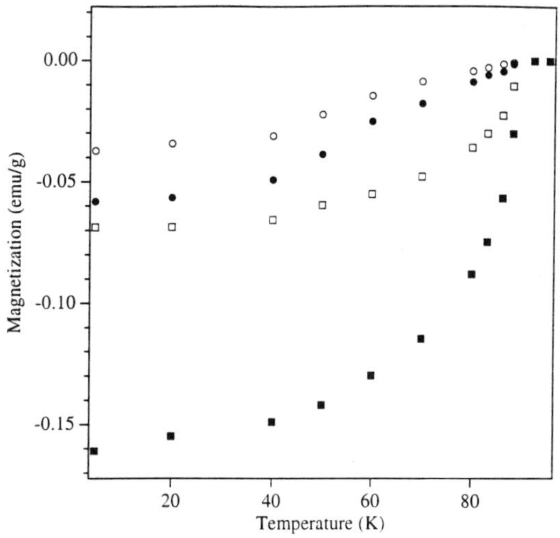

Fig. 5 Variation of magnetization with temperature for samples A (circles) and C (squares). The closed and open symbols refer, respectively, to zero-field cooled and field-cooled data.

ACKNOWLEDGMENT

Financial support by the Robert A. Welch Foundation grant No. F-1254 is gratefully acknowledged.

REFERENCES

1. A. Manthiram, J. S. Swinnea, Z. T. Sui, H. Steinfink and J. B. Goodenough, "The Influence of Oxygen Variation on the Crystal Structure and Phase Composition of the Superconductor $YBa_2Cu_3O_{7-x}$," J. Amer. Chem. Soc. **109** 6667-6669 (1987).
2. J. D. Jorgensen, B. W. Veal, A. P. Paulikas. L. J. Nowicki, G. W. Crabtree, H. Claus and W. K. Kwok, "Structural Properties of Oxygen Deficient $YBa_2Cu_3O_{7-\delta}$," Phys. Rev. B **41**, 1861877 (1990).
3. R. J. Cava, B. Batlogg, C. H. Chen, E. A. Rietman, S. M. Zahurak and D. Weider, "Single Phase 60 K Bulk Superconductor in Annealed $Ba_2YCu_3O_{7-\delta}$ $(0.3 < \delta < 0.4)$ with Correlated Oxygen Vacancies in the Cu-O Chains," Phys. Rev. B **36**, 5719-5722 (1987).
4. J. B. Goodenough and A. Manthiram, "Crystal Chemistry and Superconductivity of the Copper Oxides," J. Solid State Chem. **88**, 115-139 (1990).
5. X. X. Tang, A. Manthiram and J. B. Goodenough, "Evidence for Peroxide Formation in Superconducting $YBa_{2-x}La_xCu_3O_{7\pm\delta}$," Phys. Rev. B **37** 3734-3737 (1988).
6. A. Manthiram and J. B. Goodenough, "Factors Influencing T_c in 123 Copper Oxide Superconductors," Physica C **159**, 760-768 (1989).
7. A. Manthiram and J. B. Goodenough, "Synthesis of the High-T_c Superconductor $YBa_2Cu_3O_{7-\delta}$ in Small Particle Size," Nature **329**, 701-702 (1987).
8. H. S. Horowitz, S. J. McLain, A. W. Sleight, J. D. Druliner, P. L. Gai, M. J. VanKavelaar, J. J. Wagner, B. D. Biggs and S. J. Poon, "Submicron Superconducting $YBa_2Cu_3O_{6+x}$ Particles Made by a Low Temperature Synthetic Route," Science **243**, 66-69 (1989).
9. M. Kakihana, L. Borjesson, S. Erikson, P. Svedlindh and P. Norling, "Synthesis of Highly Pure $YBa_2Cu_3O_{7-\delta}$ Superconductors Using a Colloidal Processing Technique," Physica C **162**, 931-932 (1989).
10. J. D. Jorgensen, A. Manthiram and J. B. Goodenough, unpublished results, 1989.
11. C. Greaves and P. R. Slater, "Synthesis and Structure of the Tetragonal $Ba_3YCu_2O_x$ Phase: A perovskite Containing Carbonate Anions," Physica C **175**, 172-178 (1991).
12. Ph. Boullay, B. Domenges, M. Hervieu and B. Raveau, "Substitution of CO_3 Groups for Copper in YBCO: The Oxycarbonates $Y_nBa_{2n}Cu_{3n-1}CO_3O_{7n-3}$," Chem. Mat. **5**, 1683-1686 (1993).
13. A. Manthiram, unpublished results.

Films and Applications

Impact of HTSC-based Components on Cellular Telecommunications System Performance

H. Muller & J.D. Hodge
Illinois Superconductor Corporation
451 Kingston Court
Mt. Prospect, Illinois 60056

INTRODUCTION

Since the discovery of high temperature superconductors (HTSC) in 1987, a great deal of effort has been directed at the application of this new class of materials to commercial and military systems. While the fabrication of viable HTSC wires remains a key objective for the utilization of these materials in power and utility applications, very promising near-term opportunities exist for the use of these materials in low and intermediate frequency RF applications. These applications are attractive in that they require relatively small, discrete components that can be easily fabricated and, for the most part, do not place excessive performance demands on the existing generation of materials.

The wireless telecommunications industry is ideally suited to the application of high performance, HTSC-based components. The rapid growth in this industry, which includes traditional cellular operators (800-900 MHz) and the PCS systems currently being started up (1.8-2.0 GHz), has created opportunities for such components that are not available in other commercial RF applications. In the system architecture that is common to both cellular and PCS systems, an operating area is divided up into a number of "cells". The number and size of these cells is determined largely by the communications traffic density expected in that area. Each cell is served by a central base station which handles both uplinks and downlinks with the individual users within that cell. The base station, in turn, is in communication with a mobile telephone switching office (MTSO), which provides the interface with the "hard-wire" telephone network and controls the hand-off of a call when the user moves from one cell into another.

When a user attempts to make a call, the MTSO, by monitoring the user's signal strength at nearby base stations, assigns a base station and a channel number to the user. The selected base station communicates the channel number to the user's handset, which tunes to the appropriate frequency and the call commences. As the user moves away from the base station and the user's signal strength reaches a critical level, the MTSO polls the user's signal strength at nearby base stations and selects a new base station. The new base station is then assigned a new channel number for the conversation and the user's handset is subsequently instructed to tune to the new frequency. This hand-off typically takes place unnoticed by the user.

The FCC allocation of the cellular spectrum allows for two independent operators in every metropolitan area. The spectrum allocated for cellular use is distributed such that each operator has their spectrum interleaved with their competitor's. In addition, immediately adjacent to the cellular spectrum lies spectrum that has been allocated to special mobile radio (SMR) and air-to-ground telephone (AGT). As the popularity of wireless communications has increased, this adjacency has created significant interference problems, especially in urban environments where the bulk of wireless communications is concentrated. Cellular communications are particularly sensitive to interference, as one of it's most critical operations takes place when the user is at the periphery of the cell and, consequently, the handset signal is weakest.

Because of this sensitivity to interference, essentially all cellular base stations utilize a preselector filter. The preselector filter nominally passes only that section of the spectrum allocated to that particular operator. However, using conventional materials, filter design is a trade-off between selectivity and insertion loss. Highly selective filters can, in principle, be constructed, but only at the expense of a greatly increased insertion loss. Conversely, low loss filters can be fabricated, but only by allowing significant out-of-band interference to pass into the radio. The use of high temperature superconductors virtually eliminates the need for this design trade-off. The extremely low RF losses intrinsic to these materials enable the design and fabrication of highly selective filters that still retain a very low insertion loss. It is for this reason that HTSC-based filters represent an attractive solution to out-of-band interference problems encountered by cellular operators.

This paper describes the approach taken by Illinois Superconductor in its development of filters suitable for wireless telecommunications applications. Fundamental to this development is the focus on the use of thick film technology and three-dimensional cavity/waveguide resonant structures. As will be shown, this design paradigm enables the cost-effective, high performance filters required by the wireless industry.

HTSC THICK FILMS

Because of its ease of fabrication and attractive surface resistance at liquid nitrogen temperatures, $YBa_2Cu_3O_{7-x}$ (YBCO) was selected to develop the RF structures to be described. To achieve a sufficiently low surface resistance and RF field dependence, the microstructure of the YBCO film must be oriented. At present, two techniques are available. The first is a variant of the conventional melt-texturing process where YBCO is heated above it's peritectic temperature to decompose it to a mixture of Y_2BaCuO_5 (211) and liquid and then slowly cooled to recrystallize the YBCO. The extension of this technique to produce textured thick films has been discussed elsewhere[1,2]. Note that due to reaction with the substrate, typically zirconia, more rapid cooling rates are used to 'melt-texture' thick films than are used in bulk samples. As a consequence, textured thick films tend to develop a spherulitic growth habit. Although the presence of spherulites results in a relative low DC transport current, the behavior of these films in RF fields is acceptable, as will be shown.

The second technique for producing a textured YBCO film is via a 'reactive texturing' technique[3,4]. In this technique, a YBCO precursor is rapidly brought to a temperature and oxygen partial pressure regime where YBCO is the thermodynamically stable phase. A textured YBCO microstructure can subsequently be crystallized out of the metastable liquid that forms as a result of the rapid change of temperature and P_{O2}. The advantage of this process is that it enables the fabrication of textured YBCO films at temperatures substantially below the peritectic point. As a result, a much wider variety of substrate materials are available, including silver-plated stainless steel. Figure 1 shows an example of such a film formed on a silver-plated stainless steel substrate.

RF Characterization of Textured Films

The key parameters used to determine the quality of a given film are the surface resistance of the film and the sensitivity of the film's surface resistance to RF magnetic field ($\partial R_s/\partial H_{RF}$). For the results discussed below, these parameters were determined using a dielectric post resonator configuration, which is described in detail elsewhere[5]. Such a configuration utilizes a relatively small, flat film sample and is, consequently, an ideal resonator for routine RF measurements on HTSC samples.

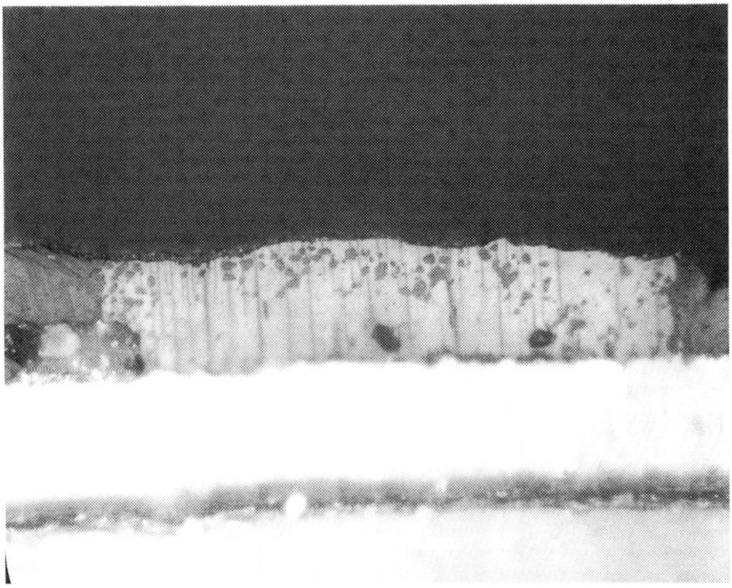

Figure 1: Microstructure of reactively textured YBCO thick film on silver-plated stainless steel. For scale, the silver layer is 50μm thick.

Figure 2 shows typical surface resistance versus surface magnetic field (roughly proportional to input power) for films produced using both peritectic recrystallization and reactive texturing. The data shown was taken at a frequency of 5.5 GHz and a temperature of 77 K using a sapphire dielectric. The shape of these curves is characteristic of what is observed for most high temperature superconductor films. At low power (RF magnetic field), the surface resistance of the film is independent of field. As the input power is increased passed a field level that is roughly dependent upon the degree of texture in the material, the material R_s begins to degrade rapidly. Note that the data shown is taken at 5.5 GHz, but that these values can readily be extrapolated to wireless frequencies using a $1/f^2$ scaling, which has been shown to be accurate in this range of frequencies.[6]

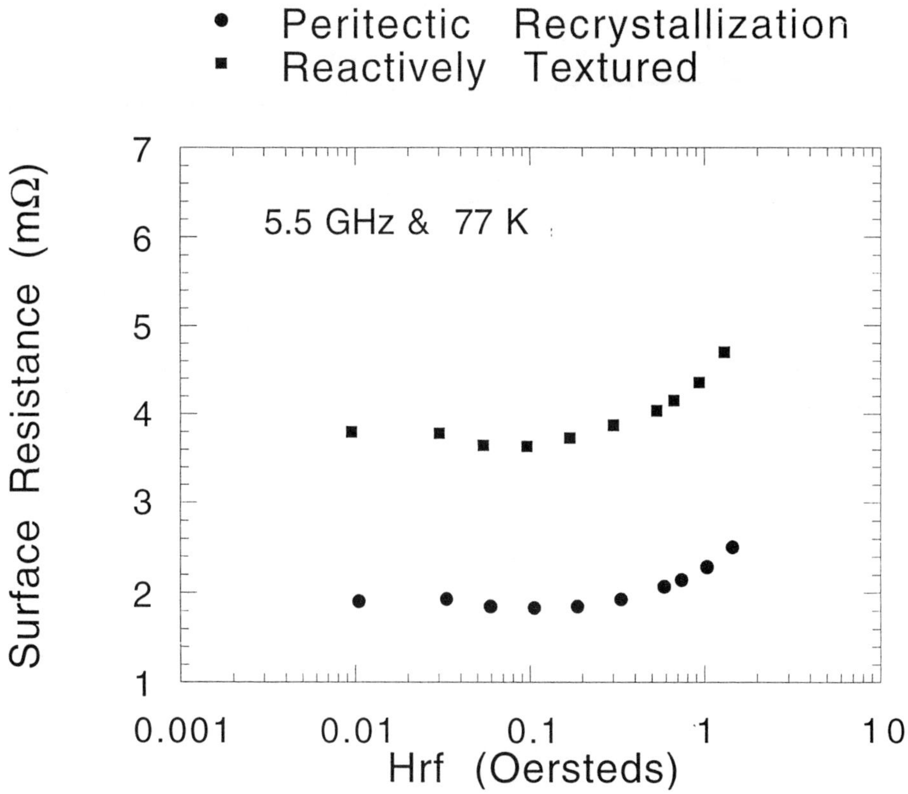

Figure 2: Dependence of surface resistance on RF magnetic field for conventionally 'melt-textured' and reactively textured YBCO thick films.

HTSC-BASED RF FILTERS

Because the losses in a superconductor are field dependent, some care must be taken to avoid operation of HTSC filter elements in the relatively lossy, high field regime. If the filter elements exhibit appreciable $\partial R_s/\partial H_{RF}$, the filter will have a non-linear relation between input and output, which will cause distortion of the output signal. Such intermodulation (IM) distortion in a preselector filter is very detrimental to overall system performance, and must be avoided. Fortunately, it is possible to keep surface magnetic fields to a safe level by selecting appropriate resonator shapes and avoiding sharp edges or discontinuities which enhance local fields. When properly implemented, HTSC filter designs exhibit no more IM distortion than conventional filters.

The main advantage of using superconducting filter elements is that, due to their extremely low RF loss, it is possible to produce designs with much lower loss and higher selectivity than the best traditional products. Figure 3 illustrates this improvement by comparing performance for an ISC cellular B band filter and a high selectivity conventional filter. Outside the band, the ISC filter attenuation is significantly higher, and provides better protection from interfering signals. Within the band, the attenuation of the ISC filter is less, so desired signals appear nearly twice as strong as they would if the conventional filter was used. The notch filter within the main band, needed to suppress interference from the interleaved A band service, is sharper, and has higher rejection.

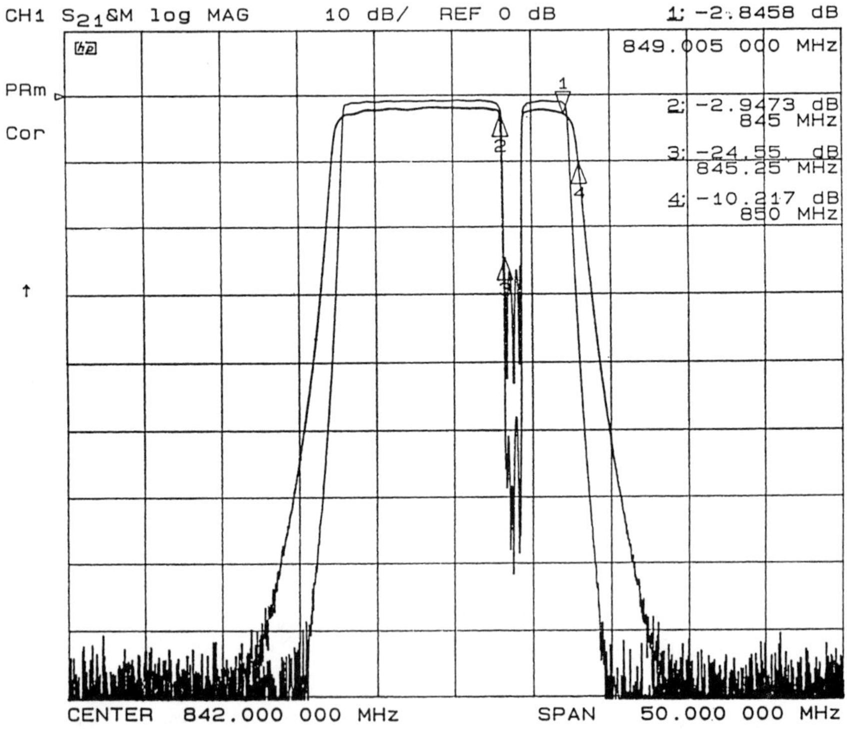

Figure 3: Comparison of ISC B band filter to a high selectivity conventional filter

FIELD TRIAL RESULTS

Over the last 18 months, ISC's superconducting filters for the cellular A and B bands have been tested in service trials in more than 13 metropolitan areas in the United States. They were undertaken to demonstrate that the selectivity and insertion loss improvements realized through the use of superconducting filters had measurable and beneficial effects on cell site operations

One early trial site was selected because it's performance was seriously affected by adjacent channel interference. It was an urban site located near a SMR transmitter, and this relatively strong signal at a frequency very close to the cellular band was causing profound interference. The main symptoms were a noticeably higher lost call rate than for other cell sites, and a higher noise floor within the cellular band. Installation of the improved filters eliminated measurable out of band signals at the receiver input, but more importantly, this resulted in operational improvements as well. The bit error rate of the site was reduced by 15%, primarily due to a lower in-band noise level. Subjective call quality was improved as well, and in comparative drive tests, highest quality connections occurred 17% more often after the filter upgrade. Finally, the number of lost calls decreased by 21%.

Another trial was conducted at a site located close to a major metropolitan airport. As in the previous trial, there was strong adjacent channel interference, in this case, from an AGT transmitter. Figure 4 shows a spectrum analyzer plot which illustrates the amount of RF energy which was present at the input of the site radios when conventional filters were used. Note that some out of band signals present were 50 dB stronger than the desired in-band signals. Figure 5 is a plot taken at the same site after the ISC filter was installed. None of the out of band signals were present, due to the improved selectivity of the HTSC filter.

After the filter upgrade, the level of in-band noise at the site was reduced by 5 dB, which caused the number of lost calls to drop by 16%. There was also substantial improvement in call quality in two areas which had enjoyed only marginal coverage prior to the trial.

A third site chosen for trial was an urban site located at the top of a 20 story building. The site was an "antenna farm', with 39 UHF transmitters sharing the same rooftop, and the RF interference was strong and widespread. The unwanted RF energy present at the site's high gain preamps was desensitizing the receiver, and so curtailing useful range. This was coupled by the ongoing switch from mobile phones, with 3 watt transmitters, to portables, with 0.6 watt transmitters. The site was not working reliably with lower power customers, and the operator

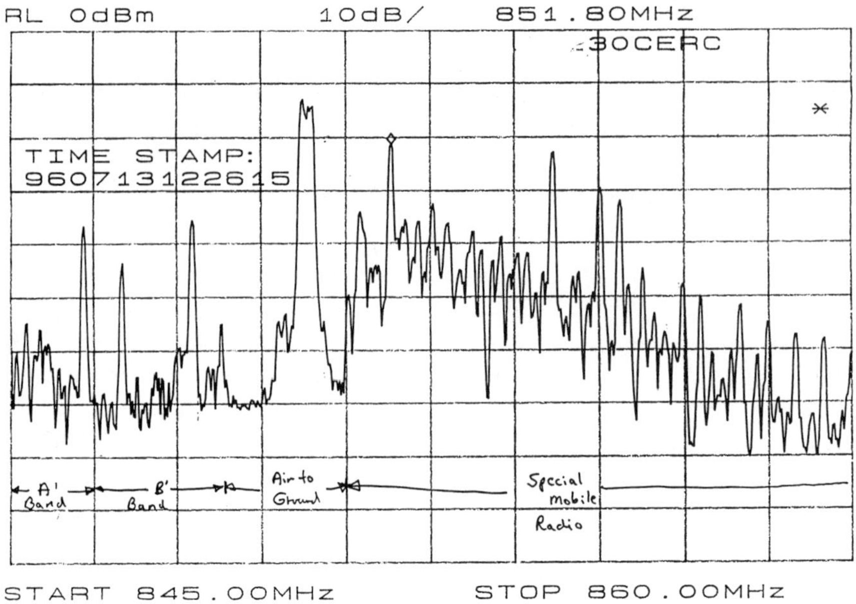

Figure 4: Spectrum at receiver input using conventional filter

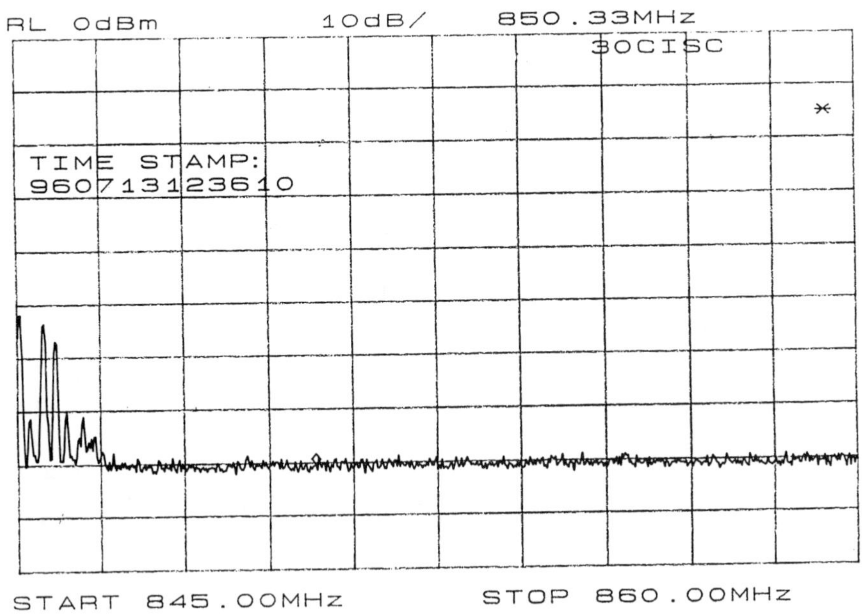

Figure 5: Spectrum at receiver input using superconducting filter

was considering installing 3 additional sites, at a cost of $2.5 million, to maintain coverage. After installing the improved filters, the downtown area accessible by portable phone was increased by 20%, and the call volume for that rooftop site increased by 20% as well.

CONCLUSIONS

Highly selective, low loss RF filters fabricated using high temperature superconductors can be very advantageously used in wireless telecommunications applications. In achieving the optimal performance of these filters, it is important to focus on both the improvement and optimization of the RF properties of superconductors and the selection of resonator geometries that accommodate the sensitivity of these materials to high RF fields.

ACKNOWLEDGMENTS

The work and results described in this paper are the result of the cumulative efforts of the the entire ISC technical and marketing staff.

---

[1] D.N. Matthews, N. Sakai, M. Murakami, and G.J. Russell, "Grain growth and seeding of thick films of $YBa_2Cu_3O_{7-\delta}$ on yttria stabilised zirconia," *Physica C.* **243** 342-353 (1995).

[2] J.S. Abell, T.C. Shields, F. Wellhofer, K.N.R. Taylor, and D. Holland, "High critical currents in thick films of YBCO showing spherulitic growth habits." *Physica C*, **162-164** 1265-1266 (1989).

[3] J.D. Hodge and L.J. Klemptner, "Superconducting 123YBaCu-Oxide produced at low temperatures," U.S. Pat. No. 5,340,797, Aug. 23, 1994.

[4] J.D. Hodge and L.J. Klemptner, "Superconducting $YBa_2Cu_3O_{7-x}$ produced at low temperatures," U.S. Pat. No. 5,527,765, Jun. 18, 1996.

[5] Z.Y. Shen, C. Wilker, P. Pang, W.L. Holstein, D. Face, and D.J. Kountz, "High T_c Superconductor-Sapphire Microwave Resonator with Extremely High Q-Values up to 90 K," *IEEE Trans. Microwave Theory and Techniques,* **40**[12] 2424-2431 (1992).

[6] J.R. Delayen and C.L. Bohn, "Temperature, frequency, and rf field dependence of the surface resistance of polycrystalline $YBa_2Cu_3O_{7-x}$," *Phys. Rev. B*, **40**[7] 5151-5154 (1989).

NORMALIZED ION POLARIZABILITY ADDITIVITY RULE AND ITS APPLICATION ON HTSC SUBSTRATE MATERIALS

Ruyan Guo and Amar S. Bhalla
Materials Research Laboratory, The Pennsylvania State University
University Park, PA 16802

ABSTRACT

An ion polarizability additivity rule was applied to calculate the dielectric constants of a large family of dielectric substrate materials and to compare with the measured dielectric constant values. The applicability and the limitations of the model were examined in light of results of new substrate materials. Preliminary set of "normalized" (to the electrostatic valence of the ion) ion polarizabilities are deduced from well established crystal structure data, taking into consideration of the apparent bonding strength. By using the normalized ion polarizabilities deduced from this approach, the applicability of the ion polarizability additivity rule is improved markedly, particularly for compounds containing transition metal elements, alkali earth elements and rare-earth elements.

INTRODUCTION

Dielectric constant of substrates for HTSC thin films is another important selection criterion in addition to the suitability of the substrate for epitaxy. To be used for microwave resonator applications where the substrates act as passive dielectrics, the suitable dielectric constant κ' values should be moderately high and the dielectric loss tangent must be low ($<10^{-4}$). To be used in the integrated circuits as the dielectric plane between the patterning, the κ' must be low (~10 or less) in order to satisfy impedance matching and to accommodate the HTSC thin films. For the ionic crystalline materials of interest, the origin of the intrinsic dielectric polarizability and the loss in the microwave region (~10 GHz) is attributed to the ionic (that includes both the ionic relaxation polarization which is temperature dependent and primarily, the tail of ionic displacive polarization from the infrared region) and the electronic polarization contributions. Theoretical calculations of the dielectric polarizability in ionic crystals require explicit information of the local electric filed that demands full knowledge on micro-structures and resolutions of the interactions among surrounding ions. Such calculation is not only complicated but also impractical as no successful local filed model has been developed in general.

The macroscopic dielectric constant and the polarizability of a molecule are known to be related through the Clausius-Mosotti relation:

$$\alpha_D = \frac{3}{4\pi} V_m \frac{\kappa' - 1}{\kappa' + 2} \qquad (1)$$

where α_D is the dielectric polarizability and V_m is the molar volume in $Å^3$. Though there has been no major breakthroughs in local field calculation for complex ionic solids, there has been steadily increasing pool of experimentally measured dielectric constants of substances of various compositions and structures. Dielectric polarizabilities and hence the dielectric constants of new materials and compounds whose dielectric constants have not been measured are potentially predictable by linear addition of the molecular polarizabilities of simpler substances (molecular polarizability additivity rule)[1,2,3] or ion polarizabilities of individual ions (ion polarizability additivity rule).[4,3] Review and comments about the application of polarizability additivity rules can be found in Shannon's paper.[5]

Molecular polarizabilities were used to calculate the dielectric constants of several substrate materials ($CaYAlO_4$, $CaNdAlO_4$, and $SrLaAlO_4$) of K_2NiF_4 structure and the comparison between the calculated and observed polarizabilities shown to have good agreement.[6] Shannon[5] has derived a set of 61 ion dielectric polarizabilities from a least square refinement procedure using the ion polarizability additivity rule and calculated dielectric constants for about a hundred compounds utilizing the Clausius-Mosotti equation. Excellent agreement between the polarizability values calculated and those experimentally measured was shown for many ternary systems including borates, aluminates, gallates, silicates, germanates, phosphates, and vanadates.

Earlier we have proposed numbers of materials for potential substrate applications.[7] Complex perovskite compounds such as $Sr(Al_{1/2}Ta_{1/2})O_3$ and $Sr(Al_{1/2}Nb_{1/2})O_3$,[8] $Ba(Mg_{2/3}Ta_{1/3})O_3$,[9] $La(Mg_{1/2}Ti_{1/2})O_3$,[10] and a family of materials of the magnetoplumbite structures[11] have been synthesized and their dielectric properties studied. It is of great interest for us to apply the ion polarizability additivity rule to a family of substrates or potential substrate materials for those we have gathered fair amount of dielectric and structural data that are previously unavailable. It is especially interesting to examine if the ion polarizability data deduced by Shannon can be used to predict the dielectric constant of new materials and to identify the "abnormal" behaviors in a material if present.

In this paper we first report the application of additivity using available ion polarizabilities to the substrate materials, next discuss the discrepancies between the experimental and the predicted dielectric polarizabilities (and the dielectric constants); then describe a new approach applying apparent bond valence and its correlation to the bonding distance to derive the "universal" or "normalized" (to its electrostatic bond valence) ionic polarizability; finally, give some examples comparing the experimental results with the predicted values obtained using the preliminary set of normalized ionic polarizabilities. The potential applications of the normalized ionic polarizabilities and ion polarizability additivity rule are shown to be promising and extension of this work is proposed.

APPLICATION OF THE POLARIZABILITY ADDITIVITY MODEL TO THE SUBSTRATE MATERIALS

Shannon's Ion Polarizability and The Additivity Model

The method to obtain the ion polarizability data and the basic approach of the application of them to predict the dielectric constants described by Shannon are schematically summarized in Fig. 1. Several assumptions were applied in the approach made by Shannon that include:

- Polarizabilities of cations and fluorine are independent of the compound in which the ions are found.
- Polarizability of oxygen was assumed to depend exponentially on the volume occupied by an oxide ion V_{OX}. This dependence can be disregarded for $V_{OX}<25Å^3$.
- Polarizability of B^{3+} was chosen to be $0.05Å^3$ that gave the polarizability of O^{2-} to be $2.01Å^3$.

Comparison of Experimental and Predicted Values

The calculation results obtained using Shannon's ion polarizability data (Table III, second column in ref. 5) and the ion polarizability additivity rule are shown in Table I. Comparison of the calculated and experimental molar polarizability and dielectric constants are tabulated. All the materials chosen for this study were substrates which have been considered for use for HTSC film deposition or have been reported as effective substrate for HTSC thin films. Materials which

ION POLARIZABILITY ADDITIVITY MODEL: SCHEMATICS USED IN THE APPROACH
by R. D. Shannon

DATABASE (129 Oxides and 25 Fluorides)

⇩

Least Squares Refinement+Clausius-Mosotti Equation:

$$\alpha_D = \frac{3}{4\pi} V_m \frac{\kappa' - 1}{\kappa' + 2}$$

⇩

Empirical Ion Dielectric Polarizabilities (61 ions)

⇩

Assumption: Total Molar Dielectric Polarizability of a Compound α_{cal}=Simple Linear Combination of Individual Ion Dielectric Polarizabilities α_i

$$\alpha_{cal} = \sum_{i=1}^{N} n_i \alpha_i$$

⇩

To Estimate Mean Dielectric Constants of "Well-Behaved"
Compounds (Also To Distinguish Unusual Dielectric Behavior).

$$\kappa_{cal} = \left(\frac{3V_m}{4\pi\alpha_{cal}} + 2\right) \Big/ \left(\frac{3V_m}{4\pi\alpha_{cal}} - 1\right)$$

Fig. 1. Schematics used to deduce the ion polarizabilities by R.D. Shannon.[5]

Table I. Comparison of calculated and experimental dielectric constant using ion polarizability value in ref. 5.

Substance	Sym.	V_m	κ'_{exp}	κ'_{cal}	$\Delta\alpha\%$
Al_2O_3	R3ch	42.45	9.34, 11.54	10.05	0.06
CaF_2	Fm3m	40.75	6.81	6.77	0.23
MgO	Fm3m	18.68	9.65	9.86	-0.61
Ba_2YSbO_6	Cubic	139.53	25.00	23.71	0.63
$Zr_{0.935}Y_{0.065}O_{1.968}$	P42/nmcz	33.65	26.00	28.50	-0.98
$Sr(Ga_{.5}Ta_{.5})O_3$	Cubic	61.58	26.90	31.54	-1.60
$CaGa_{12}O_{19}$	Hex	340.00	9.70	9.16	1.65
MgF_2	P42/mnm	32.40	5.50	5.31	1.73
$Sr_{0.5}Ca_{0.5})(Ga_{0.5}Nb_{0.5})O_3$	Cubic	58.64	32.00	25.41	2.32
SiO_2 (stru.quartz syn)	P6222	39.37	4.42, 4.41, 4.6	4.26	3.10
$BaZrO_3$	P23	73.10	43.00	27.59	3.72
$NdGaO_3$	Pbn21	57.68	20-25	31.66	-3.80
$CaYAlO_4$	Tetra	79.85	21.44, 16.12	15.54	3.80
$Sr(Al_{.5}Nb_{.5})O_3$-SAT-NG	Cubic	63.30	22.80	17.13	4.08
$Nd_{.39}Sr_{.61}Al_{.695}Ta_{.305}O_3$		56.76	24.04	39.20	-4.79
$CaGa_6Al_6O_{19}$	Hex	285.00	18.20	13.77	4.90
$La(MgAl_{11})O_{19}$	Hex	296.86	13.50	10.81	5.03
$GdAlO_3$		51.87	19.50	29.17	-5.03
$LiBaF_3$	Cubic	63.40	11.71	14.98	-5.40
$Sr(Al_{.5}Nb_{.5})O_3$-0.3$NdGaO_3$	Cubic	63.30	23.50	16.06	5.49
$Sr(Al_{.5}Nb_{.5})O_3$-0.3$LaAlO_3$	Cubic	63.28	25.90	17.02	5.63
$SrLaAlO_4$	Tetra	88.99	16.70, 20.02	28.30	-6.19
$Sr(Al_{.5}Nb_{.5})O_3$	Cubic	59.09	17.30	27.08	-6.19
$KMgF_3$	Pm3m	63.00	5.98	6.97	-6.66
$KZnF_3$	Pm3m	66.70	8.85	7.20	6.85
$Sr(Al_{.5}Ta_{.5})O_3$-0.3$LaAlO_3$	Cubic	58.08	21.90	45.52	-7.08
$La(Mg_{2/3}Ta_{1/3})O_3$	Hex	74.33	24.10	14.56	7.30
$Nd(MgGaAl_{10})O_{19}$	Hex	296.50	14.90	10.60	7.37
$Sr(Al_{.5}Ta_{.5})O_3$-0.3$LaAlO_3$	cubic	57.89	21.90	47.94	-7.44
$PrAlO_3$	R3-mr	53.25	25.00	64.81	-7.45
$SmAlO_3$	Pbnm	52.30	19.00	38.53	-8.03
$BaSnO_3$	P23	69.40	18.00	36.06	-8.37
$LaGaO_3$	GdFeO3	59.10	25.00	81.33	-8.45
$Ba_{0.8}Sr_{0.2})(Mg_{1/3}Ta_{2/3})O_3$	Cubic	67.48	25.90	91.67	-8.46
$(Ba_{0.9}Sr_{0.1})(Mg_{1/3}Ta_{2/3})O_3$	Cubic	67.84	25.30	123.65	-9.66
$Ba(Mg_{1/3}Ta_{2/3})O_3$	Hex	68.59	24.60	138.91	-10.31
$La(Mg_{.5}Ti_{.5})O_3$	cubic	60.10	26.50	355.40	-10.83
$NdALO_3$	R3ch	52.80	17.50	46.89	-10.93
$Sr(Al_{.5}Ta_{.5})O_3$-0.3$NdGaO_3$	Cubic	57.90	16.30	42.22	-11.49
$LaAlO_3$	R3mr	54.40	21.90	406.86	-13.51
$YAlO_3$	P63/mmc	61.69	16.83, 15.94	8.79	14.13
$Sr(Al_{.5}Ta_{.5})O_3$ ordered	Cubic	59.09	11.78	37.36	-18.09

appeared in Shannon's database or calculation are also included for the sake of completeness. Most of the structural data, symmetry parameters and the molar volume (V_m), are reasonably well established and the refined single crystal structure information used were found in '*The Inorganic Crystal Structure Database (ICSD)*'.[12] Crystal structures and the molar volumes of new substrate materials are determined on available single crystal samples (grown by a laser heated pedestal growth technique) by x-ray diffraction. Ceramic samples of new substrate materials are also used, only for those well sintered samples having high density and cubic or pseudocubic symmetries. The experimental value of the dielectric constants were taken from Shannon's database wherever available. Dielectric constants of new substrate materials are measured using a high precision capacitance measurement system (General Radio 1621) in frequency range 10^3-10^5 Hz at room temperature and resonance technique or cavity perturbation techniques at the microwave frequencies (GHz). Individual details regarding the synthesis, preparation, and properties of new substrate materials may be found in referred publications.

Several observations on the results listed in Table I are to be pointed out for further discussions. First, it is strikingly apparent that even though there are small discrepancies in molar polarizability, large discrepancies in dielectric constant prediction are often found that indicates that the relation in Equation (1) may not apply. Second, good agreements are usually seen in the case of substances containing cations having high valence and small sizes, while large discrepancies are common in ternary systems that involve large cations in high coordination sites. Third, poor agreements are apparent in materials contains transition-metal ions (Nb, Ta) such as in the case of $Ba(Mg_{1/3}Ta_{2/3})O_3$, $Sr(Al_{0.5}Ta_{0.5})O_3$, and $Sr(Al_{0.5}Nb_{0.5})O_3$ and cations of rare-earth family such as in the case of $LaAlO_3$. and $LaGaO_3$.

ORIGIN OF LARGE DISCREPANCY WITH THE PREDICTED MOLAR POLARIZABILITY VALUES

Obviously any inaccuracy in dielectric constants measurements or structure parameter determinations will be the up front cause for the discrepancies. All the data used in our calculations are believed to be the most reliable data available.

Besides any unavoidable inaccuracies in the experimental results, large deviations from additivity rule using Shannon's ion polarizability data can sometimes be traced to, as pointed out by Shannon, unusual properties of the compounds such as ionic or electronic conductivity, the presence of dipolar impurities, or piezo- or ferroelectricity. However, above mentioned irregularities are presumably non existent in the substrate materials under current study.

After taking into consideration the oxygen polarizability dependence on the volume, V_{ox}, the calculation in Table I was repeated. The improvement is insignificant and discrepancy is largely non-corrected.

A critical question needed to be answered after examining the ion polarizability calculation is that whether the assumption is valid that cation (and fluorine) polarizability is independent of the compound in which it is found?

There seems to be enough indication that suggests the answer to this question is no. Evidence were presented by Safford and Silverman[13] that a change of coordination of an ion alters the molar refraction contribution of the ion or more correctly of the group immediately surrounding the ion. Fajans and co-workers[14] showed that the coordination number of Zn^{2+} is related to the molar volume (of Zn^{2+}) and demonstrated a relationship between molar refraction and molar volume (of the cation). These authors showed that the Al^{3+} in fourfold coordination in glass or in certain crystals has a molar refraction contribution of 12.3 (calculated for Al_2O_3), whereas in six-fold coordination it is 10.5. Kreidl[15] first indicated that

MgO may assume both "basic" and "acidic" properties in glass. Difference in molar refraction were found for Mg^{2+} in six-fold coordination such as in periclase MgO where R=4.538, and in fourfold coordination such as in spinel $MgAl_2O_4$ where R=5.18.[16] For ions preferably found only in the same coordination, the dielectric polarizability of the ion remains relatively constant and the additivity rule applies well. Molar volumes of the compound in those cases are all that one needs to calculate to obtain the total polarizabilities of the compounds. Examples of this type would be the silicates and borates. For substances that contains cations that may be found in different coordination, the difference in the ionic polarizability is averaged when enough substances are used in the least square fitting process and therefore good or fair additivity application can be obtained. Examples of this type may be identified as aluminates and gallates. However, for cations like Cu^{2+}, Ba^{2+}, Fe^{3+}, RE^{3+}, Sb^{4+}, As^{3+}, Nb^{5+}, and Ta^{5+}, the ion polarizability data were scarce and single crystal data were mostly not available, the uncertainties are large because the ratio of the number of observed molar polarizabilities to refined ion polarizabilities is rather low and the observed total polarizabilities have larger errors (as also noted by Shannon).

Another point may also need to be pointed out is that the calculation using the ion polarizability additivity rule is based on the assumptions that the crystals considered are purely ionic, that the dielectric polarizability of a crystal is simply the sum of the dielectric polarizabilities of the individual ions, and that the Lorentz factor has a normal value of $4\pi/3$. It has been suggested[17] that the normal Lorentz factor may not be correct for some crystals because of the extended nature of the electronic charge distribution and the consequent overlapping of neighboring ions. The ion additivity rule may not be suitable in the case where covalent bonding being prominent, for instance. It also can not be unambiguously used in the case of ionic crystals with low symmetries (other than cubic symmetry). All of these mentioned above point out the need of modification for obtaining a more reliable compilation of ionic polarizabilities to be used in predicting dielectric constants in ionic solids.

THE DERIVATION OF NORMALIZED ION POLARIZABILITY

Following the discussion in the previous section, if the dielectric polarizability of a cation is dependent on individual compound, can a set of generally constant ion polarizability data be found and valid for use by the additivity rule for a broad range of compounds with various coordination combinations or crystallographic symmetries? Following the work of O'Keeffe[18] and Brown and Altermatt[19] on the apparent bond valences and their relations to bond lengths, we suggest a reformulating of the ion polarizability additivity rule by normalizing polarizability of an ion to its electrostatic valence states. This new approach defines normalized ion polarizabilities that can be universally correct for estimation of the dielectric constants with higher precision for compound with known crystal structures.

Correlation of Apparent Bond Valence and Bond Length

Unlike the *bond valence*, v_i = (electrostatic valence V_i)/(coordination Z), that is a topological property of a crystal structure, *apparent bond valence* (v'_{ij}) is a property of a bond that is calculated from its length using a predetermined equation therefore an experimentally determined quantity. The sum of apparent bond valences for all the bonds to a given atom are defined as the *apparent valence* $V'_i=\Sigma_j v'_{ij}$.

The correlation of bond length, d_{ij}, with apparent bond valence, v'_{ij}, has been reviewed by Brown[20] and the expression was given:

$$v'_{ij} = exp[(R_0 - d_{ij})/b] \qquad (2)$$

where R_0 is a parameter characteristic of the atom pair *i, j,* and *b* is a second parameter. In practice, *b* can be maintained constant at b=0.37 Å for a wide variety of bonds. Value of R_0 are empirically determined and many of which for bonds from elements to oxygen have been examined and tabulated.[21,20] In what follows, the b and R_0 values recommended by Brown and Altermatt are used where available.

In this approach, atoms are considered different if they are distinct in the crystallographic sense. When a set of bond length was known, the crystal structure would either be known or severely constrained. Being affected by factors more than bond valences, the apparent valences of atoms are often significantly different from the actual (electrostatic) valences. Equation (2) therefore controls the way the remaining interatomic distances change around a given atomic center, when one particular distance is increased or decreased. It also gives a feel for the relative strengths of contacts of various types. In crystal structure determination it is the primary means to distinguish oxide and hydroxide species as well as to the determination of site preferences in mixed systems.[22] What we propose is a new application of this consideration as to derive normalized ion polarizabilities to be used in ion polarizability additivity rule.

First, substance containing the ions of interest is selected. The structure of the substance must be well characterized and refined. Bonding distances between each cation to the (bonded) surrounding oxygen are used to calculate the apparent bond valences. Summation of all the apparent bond valence around a center ion give rise to the value of apparent valence of the ion. Ratio of apparent valence to electrostatic or "true" valence, $\Gamma = V'/V$ is calculated.

The ratio of apparent valence to true valence, Γ, is then used as a modification term for the polarizability of the cation, and the total molar dielectric polarizability is expressed as the summation of all apparent ion polarizability, $\alpha_i \Gamma_i$:

$$\alpha_D = \sum_{i=1}^{N} n_i \alpha_i \Gamma_i \qquad (3)$$

The α_i derived by this way is hence termed *normalized ion polarizability* as the deviation of the apparent valence from the ideal valence, if there is any, is already corrected in the calculation. Because the apparent bond valence is calculated for each bond, ions in different coordination systems are distinguished and yield distinct ionic polarizabilities.

The ratio Γ is calculated for a number of simple oxides and their compounds and correlates generally with the trends in bond energies of the binary compounds, as seen in Table II. Γ is shown to have less than unity for the large alkaline earth elements which signifies that there are large differences in radii and electronic polarizabilities between cations and anions that cause electronic charge distribution. On the other hand Γ is close to or exceeds unity for cations of higher charge but smaller radii, which may be evidence of electronic charge overlapping among neighboring ions. In cases where Γ severely deviates from unity, it is evident that local field differ from Lorentz field and Clausius-Mosotti relation is not satisfied. It may be pointed out that it is the "apparent" ion polarizability ($\alpha'_i=\alpha_i\Gamma_i$) that were refined in Shannon's ion polarizability data. When the number of sample is small, the accuracy of the refined ion polarizability can suffer for cations that are found in various coordination or crystal structures.

The current normalized ion polarizability additivity rule extracts normalized ion polarizability from the refinement database, taking into consideration the coordination system and making allowance for deviation of Clausius-Mosotti relation by correction. This modification has another advantage as the number of samples (database) does not need to be large. As long as the dielectric property and crystallographic structure of the substance used as a database entry for refinement are precisely known, the deduced ion polarizability is to be expected to have good reliability since only the normalized contribution of the ion is taking into account regardless the specific crystallographic surroundings the ions are involved.

Table II. Examples of the ratio of apparent valence to electrostatic valence

Compound	Element (CN):	Γ=V'/V
Li2O	Li IV:0.9395	
LiF	Li VI:1.0373	
BeO	Be IV:0.9700	
NaF	Na VI:1.0955	
KF	K VIII:1.13359	
MgO	Mg VI:0.983	
MgF2	Mg VI:0.9944	
CaO	Ca VI:0.9313	
CaF2	Ca VIII:0.9837	
SrO	Sr VI:0.8857	
Al2O3	Al VI:0.99744	
In2O3	In VI:1.01488	
SiO2	Si IV:1.0732	
GeO2	Ge VI:1.0456	
SnO2	Sn VI:0.9942	
PbO	Pb IV:0.9949	
LaF3	La IX:0.944	
CeF3	Ce IX:0.9531	
CeO2	Ce VIII:1.213	
Eu2O3	Eu VI,VII:0.9815	
Gd2O3	Gd VI:0.7584	
Er2O3	Er VI:0.9326	
Yb2O3	Yb VI:0.9360	
TiO2	Ti VI:1.0184	
MnF2	Mn VI:0.9634	
ZnO	Zn IV:0.9451	
ZnF2	Zn VI:0.994	
Y2O3	Y VI:0.9784	
ZrO2	Zr VII:0.9588	

Compound	Element (CN):	Γ=V'/V	
ZrO2 (Hi. Pressure)	Zr VIII:0.9203		
Nb2O5	Nb IV,VI:1.014		
Nb2O5	Nb VI:0.9939		
BaB2O4	B III:0.9977	Ba VIII:1.0222	
Mg2SiO4	Mg VI:0.9794	Si IV:0.9698	
MgTiO3	Mg VI:0.68386	Ti VI:1.4449	
BeAl2O4	Al VI:0.96375	Be IV:0.977	
MgAl2O4	Al VI:0.9517	Mg IV:1.07793	
YAlO3	Al VI:0.9916	Y XII:1.0694	
LaAlO3	Al VI:1.035	La XII:1.016	
PrAlO3	Al VI:1.07413	Pr XII:0.9745	
NdAlO3	Al VI:1.04268	Nd XII:1.00672	
SmAlO3	Al VI:1.025	Sm VIII:0.9125	
KMgF3	K XII:1.3037	Mg VI:0.6580	
KMnF3	K XII:0.8739	Mn VI:1.030	
KNiF3	K XII:1.2251	Ni VI:0.9919	
KZnF3	K XII:1.1297	Zn VI:1.0038	
YVO4	V IV:1.0102	Y VIII:1.0774	
LiGaO2	Ga IV:0.9709	Li IV:0.9848	
LiGaO2 (Hi. Pressure)	Ga VI:0.9622	Li VI:0.9687	
SrWO4	Sr VIII:1.106	W IV:0.9691	
SrMoO4	Mo IV:0.9738	Sr VIII:1.10	
BaWO4	Ba VIII:1.11691	W IV:0.96125	
Li2GeO3	Li IV:0.9655	Ge IV:1.0589	
MgAlBO4	Al VI:1.0297	B IV:0.9795	Mg VI:1.0286
CaNdAlO4	Al VI:1.211	Ca IX:0.865	Nd IX:0.837
CaMgSi2O6	Ca VIII:1.0367	Mg VI:1.0539	Si IV:0.9741
BaZnGeO4	Ba 689:1.113	Ge IV:1.033	Zn IV:1.426
Gd3.00Ga5.00O12	Gd VIII:1.0078	Ga VI:0.948	Ga IV:0.969

The modified approach as described above are schematically summarized in Fig. 2. The preliminary set of normalized ion polarizability, particularly shown the comparison between different coordination, is presented in Fig. 3.

The calculated dielectric constants using the normalized ion polarizability additivity rule are listed in Table III. The improvement between the experimental value and calculated value for large alkaline-earth ions (Sr, Ba), transition metal ions (Nb, Ta), and rare earth ions (La, etc.), is apparent.

The current modification is however anything but a comprehensive treatment for a rather complicated topic. The one parameter linear correction used in deducing the normalized ion polarizability is currently a successful solution to extract more reliable data from a limited database. It nevertheless provides no solution for anisotropic properties. Like any other empirical method, the reliability of the normalized ion polarizability data depends on the quality of the database available and further improvement is anticipated.

Detailed examples and applications of the normalized ion polarizability additivity rule are discussed in depth in an upcoming publication.

ACKNOWLEDGMENT

This work was financially supported by the Defense Advanced Research Projects Agency and the Office of Naval Research.

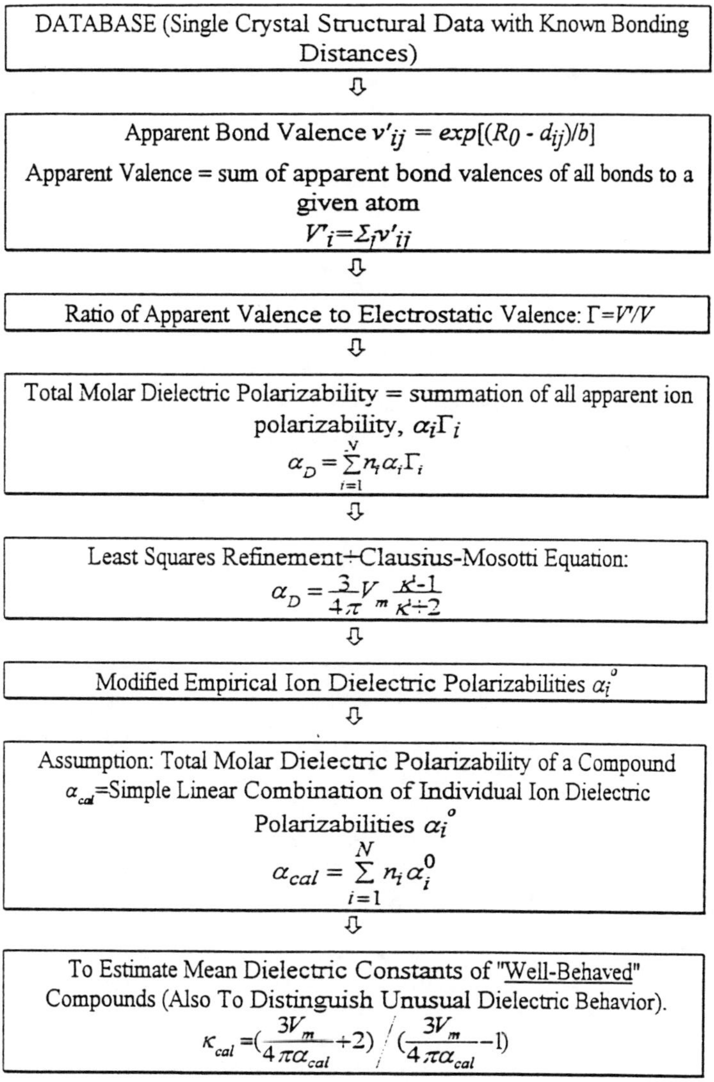

Fig. 2. Normalized ion polarizability additivity rule: schematic of the approach.

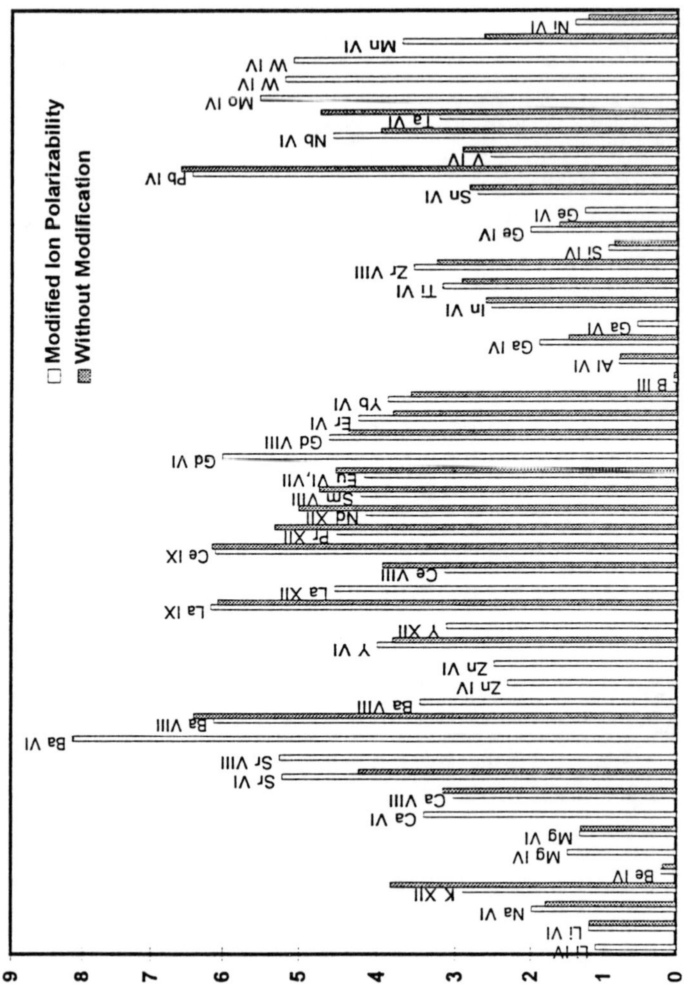

Fig. 3. Comparisons of ion polarizability given in Å^3, corresponding to respective valence and coordination.

Table III.: Examples of calculated dielectric constants using modified ion polarizability.

Substance	Sym.	V_m	κ'_{exp}	κ'_{cal}	$\Delta\alpha\%$
$Ba(Mg_{1/3}Ta_{2/3})O_3$	Hex	68.59	24.60	24.29	0.15
$(Ba_{0.8}Sr_{0.2})(Mg_{1/3}Ta_{2/3})O_3$	Cubic	67.48	25.90	21.74	2.11
$Ba_{0.9}Sr_{0.1})(Mg_{1/3}Ta_{2/3})O_3$	Cubic	67.84	25.30	23.48	0.88
$CaNdAlO_4$	Tetra	82.30	19.65,17.65	18.34	-0.29
$CaYAlO_4$	Tetra	78.88	21.44,16.12	22.44	1.97
$LaAlO_3$	Rhomb	54.40	21.90	22.39	-0.87
$Nd_{.39}Sr_{.61}Al_{.695}Ta_{.305}O_3$		56.76	24.04	22.17	1.01
$PrAlO_3$	Rhomb	53.25	25.00	25.99	0.44
$Sr(Al_{.5}Ta_{.5})O_3$ ordered	Cubic	59.09	11.78	17.10	-7.76
$Sr(Al_{.5}Ta_{.5})O_3$-0.3$LaAlO_3$	Cubic	58.08	21.90	22.75	-0.50
$Sr(Al_{.5}Ta_{.5})O_3$-0.3$NdGaO_3$	Cubic	57.90	16.30	21.89	-4.59
$SrLaAlO_4$	Tetra	89.15	16.81,20.02	18.77	0.61

REFERENCES

1 A. Heydweiller, *Z. Phys.* **3**, 308 (1920).
2 D.A.A.S. Narayana Rao, *Proc. Ind. Acad. Sci.* **30A**, 317 (1949).
3 A.C. Lasaga and R.T. Cygan, *Amer. Mineral.* **67**, 328 (1982).
4 R. Roberts, *Phys. Rev.* **76**, 1215 (1949).
5 R. D. Shannon, *J. Appl. Phys.* **73**(1), 348 (1993).
6 R.D. Shannon, R.A. Oswald, J.B. Parise, B.H.T. Chai, P. Byszewski, A. Pajaczkowska, and R. Sobolewski, *J. Solid State Chemistry* **98**, 90-98 (1992).
7 R. Guo, A.S. Bhalla, Rustum Roy and L.E. Cross, *Epitaxial Oxide Thin Films and Heterostructures*, **v.341** 215-220, Eds. Ramesh, Fork, Phillips and Wolf, MRS Publications (1994).
8 R. Guo, J. Sheen, A.S. Bhalla, F.W. Ainger, S. Erdei, E.C. Subbarao, and L.E. Cross, *J. Materials Research*, **10(1)** 18-25 (1995).
9 R. Guo, A. S. Bhalla, and L.E. Cross, *J. Appl. Phys.*, **75(9)** 4704-4708 (1994).
10 G. Harshe, A.S. Bhalla, and L.E. Cross, *Materials Letter* **18**(4), 173 (1994).
11 R. Guo and A.S. Bhalla, "Materials of Magnetoplumbite Structure for High Tc Superconductor Substrates," presented in The America Ceramic Society 95th Annual Meeting, April 18-22, 1993, Cincinnati, Ohio.
12 The Inorganic Crystal Structure Database (ICSD), copyrighted by Fachinformationszentrum-Karlsruhe, D-7514 Eggenstein-Leopoldshafen 2, Karlsruhe, West Germany.
13 H.W. Safford and A. Silverman, *J. Am. Ceram. Soc.*, **30**, 203 (1947).
14 K. Fajans and N.J. Kreidl, *J. Am. Ceram. Soc.*, **31**, 105 (1948); K. Fajans and G. Joos, *Z. Physik*, **23**, 1 (1923).
15 N.J. Kreidl, *Glastchn. Ber.*, **7**, 313 (1929).
16 Rustum Roy, *J. Am. Chem. Soc.*, **72**, 3307 (1950).
17 N.F. Mott and R.W. Gurney, *Electronic Processed in Ionic Crystals* 2nd edition, p.16 (Oxford University Press, London, 1948).
18 M. O'Keeffe, *Structure and Bonding*, **71** (Springer-Verlag, Berlin Heidelberg, 1989) p.163-190.
19 I.D. Brown and D. Altermatt, *Acta Cryst.* **B41**, 244 (1985).
20 I.D. Brown, *Structure and bonding in crystals II*, M. O'Keeffe, A. Navrotsky (eds), Academic, New York (1981).
21 I.D. Brown and K.K. Wu, *Acta Cryst.* **B32**, 1957 (1976).
22 J.K. Burdett, *Chem. Rev.* **88**, 3 (1988).

DEVELOPMENT OF CERAMIC SUPERCONDUCTORS FOR ELECTRIC POWER APPLICATIONS

U. Balachandran
Energy Technolgy Division, Argonne National Laboratory
Argonne, Illinois 60439 USA

ABSTRACT

The U.S. Department of Energy supports an applied superconductivity program entitled "Superconductivity Program for Electric Power Systems." Activities under this program are designed to help develop the high-temperature superconductor (HTS) technology that is needed for industry to proceed with the commercial development of electric power applications. Research is conducted in three categories: wire development, systems technology development, and Superconductivity Partnership Initiative (SPI). Wire development acitivities are devoted to improving the critical current density (J_c) of short-length HTS wire, whereas activities in systems technology development focus on fabrication of long-length wires, coils, and magnets. Finally, SPI activities focus on the development of prototypes that consist of a generator coil, a fault current limiter, a transmission cable, and a motor. A current overview and recent progress in the development of HTSs are outlined in this paper.

INTRODUCTION

Since the advent of superconductivity in 1911, considerable effort has been expended to raise the critical temperature (T_c) below which this phenomenon occurs. A significant breakthrough was achieved in 1986 with the discovery of the so-called "copper oxide" compounds, or high-temperature superconductors (HTSs), such as lanthanum-barium-copper-oxygen (LBCO) and lanthanum-strontium-copper-oxygen (LSCO) [1-2]. Immediately following this discovery, three additional HTS compounds, namely, yttrium-barium-copper oxide (YBCO), thallium-barium-calcium-copper oxide (TBCCO), and bismuth-strontium-calcium-copper oxide (BSCCO) were synthesized [3-5]. All of these compounds have T_c values that are well above the boiling point of liquid nitrogen (77 K). Cheap and readily available liquid nitrogen could now be used to replace the costly and difficult-to-obtain liquid helium as the refrigerant.

Potentially, HTSs can be used in motors, generators, transmission cables, fault current limiters, and in the medical and electronics industry. Because of the technical and economic incentives that these uses offer, the ensuing pace of the development of HTS materials has been fairly rapid [6]. Several prototype devices have been tested under real operating conditions and plans are underway to scale them up. The results are very encouraging inasmuch as they foster the hopes of the entire scientific community for phenomenon of superconductivity. This paper addresses current issues of the HTS industry and briefly summarizes the progress that has been made.

RESEARCH AND DEVELOPMENT OF HIGH-TEMPERATURE SUPERCONDUCTORS IN THE U.S.

High-temperature superconductor technology is considered essential for American competitiveness in the next millenium. Because they can be used to build smaller and more efficient devices, design more efficient energy transportation and storage systems, and, more importantly, because the technology is environmentally friendly, the benefits of HTSs are remarkable. The global superconductivity market is projected to be $45 billion by 2010 [7]. There is no doubt that the development of HTSs will have a significant impact on both the industry and the end user. Therefore, the U.S. Department of Energy (DOE) and Department of Defense (DOD) have been actively sponsoring research and development activities on these materials. In FY 1994, the U.S. government spent an estimated $148 million for HTS research and development. The largest contribution, $70.2 million, was contributed by DOD, followed by $49.8 million from DOE, and $23.4 million from the National Science Foundation (NSF). The objective of the various programs is to develop the technology required for the commercial development of various HTS-related applications. By offering superconducting products, U.S. industry will not only regain a major share of the global market but it will also have a more efficient and flexible energy system [8,9].

To achieve this objective, DOE has established alliances, through its Superconductive Partnership Initiative (SPI) program with national laboratories, universities, and private industries. The SPI, a flagship program in the nation's effort to commercialize HTS technologies, is funded to $12 million over 2 years, with $9 million coming from DOE and $3 million from cost sharing provided by the industry. While the involved industries will be responsible for the initial demonstration and commercialization of the technology, research at universities and government laboratories is targeted for long-term technical support and improvement. Currently, four industry teams, focusing on four specific HTS products, are participating in Phase I of the SPI program [10].

The first team is led by Martin Marietta Corp. (General Dynamics) and includes American Superconductor Corp. (ASC), Los Alamos National Laboratory (LANL), and Southern California Edison (SCE). The team is working on a $4.2 million project to develop superconducting fault current limiters, which have an estimated global market of $20 billion.

The second team, which is led by General Electric Company (GE), includes the Electric Power Research Institute (EPRI), Intermagnetics General Corp. (IGC), New York State Energy Authority, New York State Institute for Superconductivity, Niagara Mohawk Power, and Argonne, Los Alamos, and Oak Ridge National Laboratories (ANL, LANL, ORNL). The team has formed a $4.2 million partnership to develop superconducting generators that are projected to lower the cost of electricity and serve a worldwide market estimated at $30 billion.

The third SPI team is led by Reliance Electric Co., and includes ASC, Centerior Energy, EPRI, and Sandia National Laboratories. This team is working on a $3.4 million project to develop superconducting motors. Large industrial motors consume nearly 30% of the electricity generated in the U.S., so the development of a more efficient motor would provide a competitive advantage in the $300-million-per-year market [10].

The fourth team comprises Pirelli Cables North America, ASC, EPRI, and LANL. The goal of the team is to develop HTS power cable systems and introduce them to the electric power market by the year 2000. Two types of HTS cables are to be developed: the first type, retrofit cables, as the name suggests, will be drawn into existing conduits. Copper cables that are worn out or exhibit lower power capacity will be extracted and replaced by the retrofit cables. The second type, coaxial cables, will be used primarily for underground installations [11].

The biggest DOD sponsor is the Advanced Research Projects Agency (ARPA), which is spending an estimated $46 million to develop various microelectronic devices. The ARPA also supports the Consortium for Superconducting Electronics, which was formed in 1989 to conduct research and development in the same area. Participants in the consortium include IBM, AT&T, Conductus, Concurrent Technologies Inc., Lincoln Laboratories, and Massachusetts Institute of Technology. The consortium is actively involved in two applications: cellular base stations for wireless communications, and medical instruments that use superconducting quantum interference device (SQUID) magnetometers and gradiometers. Microelectronics-related applications are outlined in detail in a review article by Lubkin et al. [12]. The objective of the various programs is to accelerate the introduction of HTS technology into the commercial marketplace.

CHARACTERISTICS OF HIGH-TEMPERATURE SUPERCONDUCTORS

Most of the applications that have been envisaged for HTSs are still in their developmental stages and, so far, only a few products have been introduced into the marketplace. The reason for such a low output stems from the complexities of the HTS system [13]. Although major HTS compounds exhibit T_c values well above the boiling point of liquid nitrogen, they have both advantages and disadvantages. While TBCCO and mercury-barium-calcium-copper oxide (HBCCO) are still in the developmental stages, most applications currently use either BSCCO or YBCO. Toxicity of both thallium and mercury has posed a major stumbling block to the development of TBCCO and HBCCO. Several superconducting phases have been identified in the BSCCO homologous series, the most popular versions being $Bi_2Sr_2Ca_1Cu_2O_{8+\delta}$ (Bi-2212) with a T_c of ≈80 K, and $Bi_2Sr_2Ca_2Cu_3O_{10+\delta}$ (Bi-2223) with a T_c of ≈110 K [13]. Because the homogeneity range of the Bi-2212 phase is well known, this phase can be prepared easily over a large composition range. In contrast, little information is available about the homogeneity range of the Bi-2223 phase, which has the highest T_c in the BSCCO homologous series. Two major issues seem to complicate the development of the Bi-2223 phase; it is stable in only a very narrow temperature range and the kinetics of its formation are so slow that it is almost impossible to obtain the phase-pure material [13]. Because of this, precise control over the processing parameters is required so the desired properties can be realized.

Current transport in HTSs is controlled by intra- and intergranular critical current density (J_c). High intragranular J_c values have been achieved in HTSs. Because current flow occurs by a percolative process, intergranular J_c is the main factor that controls the overall current transport property of a polycrystalline superconductor. A grain with high intragranular J_c will not be of much use if it is poorly coupled to the adjacent grain. Poor coupling or "weak link" is more severe in YBCO than in the other HTS families. In YBCO, weak links occur because of the formation of high-angle grain boundaries. These boundaries tend to act as barriers to the flow of supercurrent. According to Dimos et al., the principal way to avoid such weak links is to texture or align the grains so that their boundaries are at a low angle [14]. Salama et al., on the other hand, have recently shown that, in melt-processed polycrystalline YBCO samples, substantial current can be transported across grain boundaries with misorientations up to 30° [15]. Briefly, J_c characteristics of HTS are closely related to the distribution of grain boundary misorientations.

Apart from phase purity and grain alignment, an even more daunting challenge is to overcome the effects of an external magnetic field. In many envisioned bulk applications, such as motors, generators, transformers, and transmission cables, the presence of a magnetic field is an integral part of the systems operations [16]. An external magnetic field that is sufficiently strong can penetrate the superconductor in discrete bundles of flux lines or vortices. As long as the flux lines remain stationary, the HTS can withstand the external magnetic field without losing its superconducting properties. However, when an electric current is applied, the flux lines move or "creep" because of Lorentz force. Flux creep also occurs when the operating temperature increases in the presence of a magnetic field. Motion of flux lines blocks current flow and thereby causes dissipation of energy [8,13,17,18]. Effort is currently focused on circumventing the problem of flux motion by adopting various metallurgical techniques, for example, precipitating secondary phases, introducing fine-scale microstructural defects such as twins and inclusions, irradiating the material to form columnar defects, etc. These microstructural inhomogeneities are expected to pin the flux lines to their respective positions. By doing so, one can limit their disruptive tendency [8,13,17-21]. Modifications that are based on smart material design approaches have produced remarkable results.

YBCO samples have now been produced that carry 1 x 10^6 A/cm^2 at 77 K, dropping to only 4 x 10^5 A/cm^2 when a magnetic field of 9 T is applied. Both values are much better than the initial results, when YBCO could carry only 10 A/cm^2 and lost all conductivity in a field of only 0.01 T [21]. Although the intrinsic flux pinning of YBCO has improved, its intergranular current transport characteristics are poor because of weak links. On the other hand, despite its inadequate flux pinning, BSCCO has been preferred for various bulk applications. The irony is that the material with poor flux pinning characteristics is the one that is being most highly developed for various applications [8,18-22].

STATUS OF HIGH-TEMPERATURE SUPERCONDUCTOR APPLICATIONS

Widespread use of superconducting technology will depend on the performance of the HTSs and the economics involved in their operation. While bulk applications, such as transmission cables, motors, generators, transformers, and magnets, require that the superconducting material be either in the form of a wire or tape, microelectronic applications involve the use of superconducting material as thin films. This section briefly summarizes the progress made by the industry in developing HTSs for various applications.

Development of HTS Wires

For successful applications in high-field magnets, transmission cables, motors, transformers, etc., it is very critical that the HTS be fabricated into long-length conductors that exhibit high J_c values and mechanical reliability. However, to process HTSs, which are brittle oxide ceramics, into a more reliable and robust form is an extremely challenging and formidable task.

Because conventional techniques, such as sinter forging, hot pressing, tape casting, and jelly roll, could not be used, the powder-in-tube (PIT) technique was developed. The groups at Vacuumschmelze and Sumitomo Electric Industry (SEI) were the first to apply the powder-in-tube (PIT) technique to HTSs [13, 23-25]. In this technique, various ingredients for the precursor powder are mixed and treated so they react with each other. The resultant powder is then packed into a high-purity silver or silver alloy tube. Silver is used because it diffuses oxygen at high temperatures and thereby allows sensitive control of oxygen partial pressure during the reaction of the superconductor [26]. The tube is then swaged and drawn through a series of dies to form a single-sheath wire ≈2 mm in diameter. The wires that are obtained are then rolled to form a tape with a final thickness of ≈0.1 mm. Multifilament conductors are fabricated by restacking monofilament wires in a larger silver tube and then subjecting it to a similar deformation sequence. The final step involves thermomechanical treatment that consists of a series of heat treatments and mechanical deformation steps [13]. During this stage, the precursor powder reacts to form the desired superconducting phases.

Tapes made by the PIT technique have a desirable geometry in which a brittle superconductor oxide is surrounded by a metallic sheath that protects the superconductor core from chemical, thermal, and mechanical abrasion. The main advantage of the PIT technique is that the metallurgical techniques employed in the fabrication of the tapes are all well established and can therefore be readily scaled up for mass production of long-length conductors. Also, texturing and grain alignment, which dictate the superconducting properties of the tapes, can be achieved very easily during the final thermomechanical treatment steps [13,23,24]. However, a complete understanding of the various parameters, such as the powder characteristics, sintering temperature, mechanical processing conditions, heat treatment time, cooling rate, etc., is essential so the desired properties can be realized. By carefully controlling these parameters, several research groups have reported J_c values of 5.4 x 10^4 A/cm^2 and 6.9 x 10^4 A/cm^2 for short-length BSCCO tapes that were subjected to a combination of uniaxial pressing and heat treatment [27,28].

Because uniaxial pressing is not amenable to the fabrication of long-length conductors, a more practical approach, such as rolling, has been adopted [13,23,24]. Companies such as IGC and ASC, in close collaboration with various universities and national laboratories, have been successful in making kilometer lengths of high-quality mono- and multifilament conductors. Figure 1 illustrates the current transport properties achieved in long-length HTSs that have been manufactured by two leading U.S. companies [7]. In collaboration with ANL, IGC reported a J_c of 1.8 x 10^4 A/cm^2 at 77 K and zero applied field for a 125-m-long monofilament conductor. Similarly, a consistent J_c of 1.2 x 10^4 A/cm^2 has also been achieved in a 1.26-km-long multifilament conductor fabricated by IGC [29]. The results indicate the substantial progress that has been made in the development of long-length conductors. However, comparison of their current transport properties with those of short-length tapes indicates that there is much room for improvement.

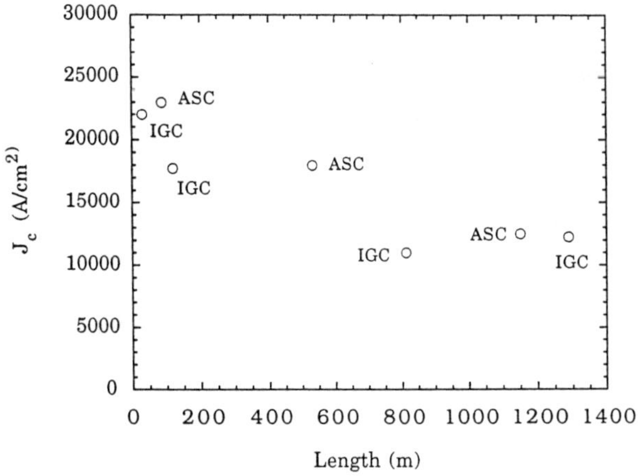

Fig. 1. Current transport properties of long-length HTSs manufactured by two leading U.S. companies.

At present, efforts are underway to develop power transmission cables from these long-length conductors. The main advantage of these cables is that, for a given cross-sectional area, the superconducting cable can carry more current, without much transmission loss, than a conventional copper or aluminum conductor. Also, it can replace the ubiquitous high-tension towers and power lines that are a constant electrical and biological hazard [26,30].

Several cable and wire manufacturing companies around the world have teamed up to develop prototype power transmission cables. A consortium consisting of the world's second largest cable manufacturer, Pirelli Cables North America, ASC, EPRI, and DOE is developing a three-phase conductor enclosed in an 8-in. steel tube. Pirelli Cable recently reported a 50-m-long prototype conductor that carries 1800 A of DC. In Japan, the Tokyo Electric Power Company (TEPCO) is collaborating with SEI and Furukawa Electric Company to develop a coaxial power cable with a transmission capacity of 0.5-1.0 GW. Recently, TEPCO and SEI conducted a 7-h transmission test with an AC of 1 kA rms on a 7-m prototype cable [8,26].

Apart from the economics of the wire development projects, a few technical issues must still be addressed. When carrying an AC or in the presence of a changing magnetic field, HTSs exhibit energy losses. These losses generate heat, which, in turn, could lead to degradation of the current-carrying properties of the tapes. The level of such losses will have a critical impact on the commercial potential and engineering feasibility of an application. By twisting the tapes and by employing an alloy sheath material, the industry plans to mitigate the AC loss characteristics of the conductors [31].

Because BSCCO exhibits poor flux pinning characteristics, applications of this superconductor have been limited to those areas where operations are conducted at 35 K and in a magnetic field of 1 T. Inasmuch as YBCO behaves better than BSCCO in the presence of a magnetic field, effort is currently underway to use it for conductor development. In 1991, Iijima and his collaborators [32] at Fujikura Limited, Tokyo successfully developed a thick film of YBCO. The film was deposited by laser ablation on a nickel alloy substrate with a buffer layer of yttria-stablized-zirconia. The group achieved a critical current density of 1.13×10^6 A/cm^2 at 77 K and zero field in short-length samples and J_c of 2.1×10^5 A/cm^2 in a 70-m-long tape. The research group at LANL [33] reported a YBCO film, ≈1-2 μm thick, that was deposited by a similar technique. This was the first time the thickness of the film had exceeded 0.5 μm. The group reported very little loss in J_c when a field of 19 T was applied parallel to the tape surface. In a field of 5 T, applied normal to the plane of the tape, J_c was 1×10^5 A/cm^2 at 77 K. These results are encouraging, because they promise to extend the range of HTS applications, especially in the presence of high magnetic fields and temperatures [33].

Recently, several research groups have reported that most of the current in the BSCCO tape flows exclusively in a thin layer of highly aligned Bi-2223 grains that are present near the silver sheath. Existence of such a layer has been confirmed by microslice experiments, high-resolution electron microscopy, and magneto-optical imaging [34-36]. A high degree of texturing and high current transport properties have been observed in BSCCO tapes that consisted of two and four coaxial layers of Bi-2223 and silver [37]. In addition, Schwartz et al. have observed improved strain tolerance characteristics in Bi-2212 tapes fabricated by the silver wire-in-tube (WIT) technique [38]. Based on these results and the ever increasing demand for conductors with high current transport properties, we tried to fabricate superconducting tapes by the WIT method. Figure 2 is an optical photomicrograph of the WIT wire we obtained. The precursor powder in this case was synthesized by the freeze-drying technique. The powder was calcined at 800°C for 2 h in a 7% oxygen atmosphere. Short samples were cut and thermomechanically processed at various temperatures between 790 and 810°C in a 7% oxygen atmosphere. Development of the WIT technique was a joint effort between the University of Pittsburgh and ANL. Figure 3 shows the variation of J_c as a function of heat treatment time at 800°C in a 7% oxygen atmosphere. The best result of $J_c > 2.0 \times 10^5$ A/cm^2 (I_c was 22 A) was obtained after 200 h and three intermediate uniaxial pressing. The results imply that the central region of the superconductor core does not contribute much to the overall current-carrying capabilities of the PIT tapes [34].

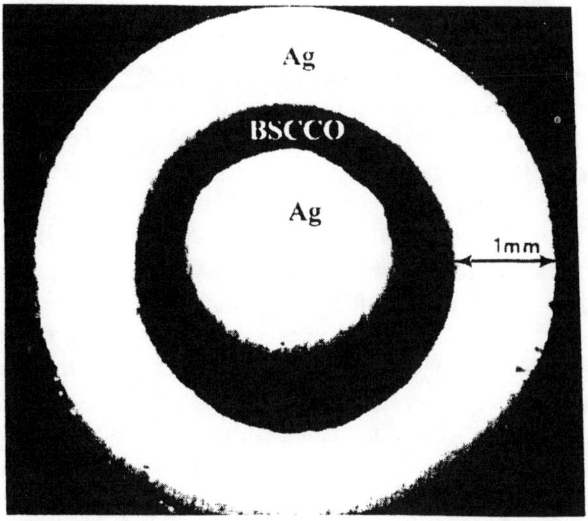

Fig. 2. Optical photomicrograph showing cross section of the silver wire-in-tube before rolling.

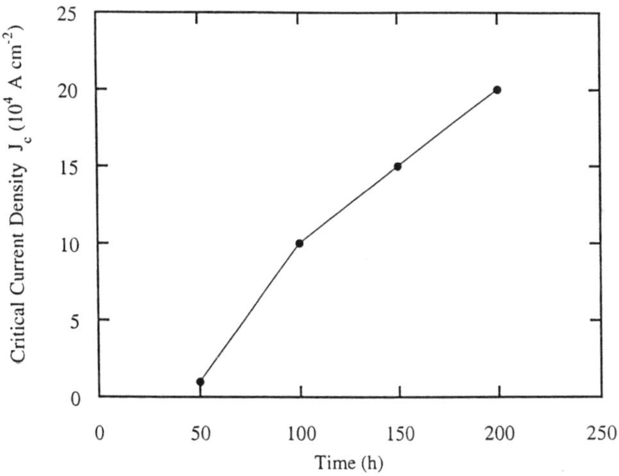

Fig. 3. Progression of J_c with time at 800°C in a 7% O_2 atmosphere; each step is accompanied by uniaxial pressing.

Poor mechanical properties have seriously hampered the commercial applications of HTSs. During fabrication and service, conductors are subjected to axial and bending stresses. The temperature gradient and magnetic field generated during operation can cause additional stresses in the material. In large and/or high-field magnets, the electromagnetic hoop stresses developed as a consequence of Lorentz force could even reach the ultimate strength of the material. These stresses can cause microstructural damage in the conductors and thereby degrade current transport properties. Although silver is widely used as a sheath material, its mechanical properties are not adequate to withstand the stresses developed during fabrication and service. Therefore, several techniques, such as adding silver to the superconductor powder, using alloy sheath material as an alternative to silver, and fabricating multifilament conductors, have been developed to improve the strain tolerance characteristics of the conductors [13, 39-41].

In-situ axial and bending tests were conducted with custom-designed test fixtures. Such in-situ characterization circumvents the errors introduced by repeated handling and thermal cycling of the tapes during external tests. Axial tests at 77 K and 0.5 T applied field indicated that the multifilament conductor exhibits better strain tolerance than the monofilament conductor, retaining >90% of its initial I_c at a strain that is ≥1%. While the irreversible strain (e_{irr}) for the multifilament conductor was ≈1%, that for the monofilament conductor was ≈0.2%. Figure 4 shows preliminary results on bending characteristics of monofilament conductors and of a multilayer composite tape, with various superconductor fill factors. The plot shows that ε_{irr} for the monofilament conductor increases with a decrease in superconductor fill factor. Also, the multilayer tape (with a fill factor of 40%) exhibited higher strain tolerance than the monofilament conductors, retaining 90% of its initial I_c at a bend strain of 1%. Thus, the added silver layer not only increases the strain tolerance of the tape, but by increasing the silver/superconductor interfacial area it should also increase the current transport properties of the tapes [42].

Magnets, Transformers, and Motors

Long-length monofilament HTSs have been co-wound to form pancake-shaped coils. A coil, wound from three 10-m lengths of rolled BSCCO tape, contained 57 turns and exhibited an I_c of ≈22 A at 77 K and 280 A at 4.2 K. The coil generated a magnetic field of ≈4200 and 330 G at 4.2 and 77 K, respectively. IGC has fabricated a superconducting test magnet by stacking such pancake-shaped coils and connecting them in series. A prototype test magnet, fabricated by stacking 20 pancake-shaped coils, with a total conductor length of 2400 m, generated a field of 3.2 T at 4.2 K and zero applied field [43]. ASC has developed a test magnet that generates a field of 0.6 T at 77 K and 3.4 T at 4.2 K. Currently, SEI holds the

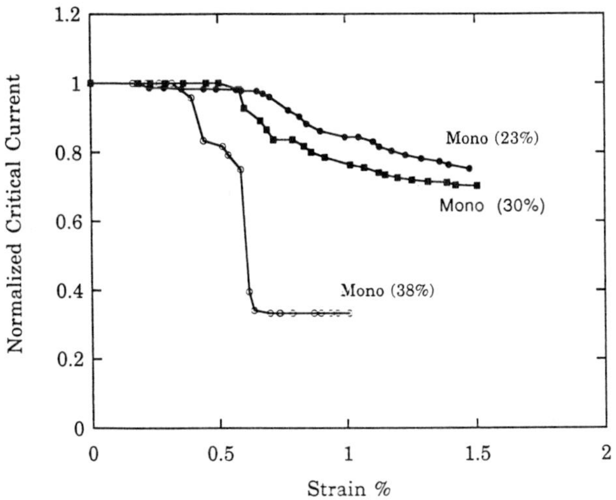

Fig. 4. Strain tolerance of silver-clad BSCCO monofilament tape with various fill factors.

record for the highest field generated by a superconducting magnet; their magnet generated a field of 4 T at4.2 K [26]. They plan to use these magnets as inserts in analytical equipment, such as that used for nuclear magnetic resonance (NMR) imaging, where low-temperature superconducting magnets are currently employed.

Intermagnetics General Corporation has developed a 0.25-KVA HTS transformer that includes a racetrack-wound solenoid. The primary end of the transformer was made of ≈85 m of BSCCO tape, the secondary end was 31 m long. Turns in the primary and secondary windings were 140 and 40, respectively. Intermagnetics General Corporation and Waukesha Electric Systems, Wisconsin have teamed with Rochester Gas and Electric and ORNL to develop a single-phase transformer rated at 1 MVA and operating at 20-50 K. The world's first prototype transformer is scheduled to be installed in the power grid for Geneva, Switzerland. The transformer is designed to produce 630 kVA at 77 K. American Superconductor Corporation will be providing the superconducting wire for ASEA Brown Boveri, which is involved in the project. Although issues such as AC loss remain, extensive studies by Pacific Northwest National Laboratory on size, weight, and cost savings indicate that transformers made with HTSs are commercially viable [8,26].

Reliance Electric Company and ASC, in cooperation with DOE, recently demonstrated a 200-hp HTS motor, which represents an output power 60% greater than the original design output of 125 hp. The 200-hp motor was designed and built by Reliance and included HTS rotor coils supplied by ASC. The motor was operated at 1800 rpm, with the rotating superconducting coils at 27 K [44]. The Naval Research Laboratory and the Naval Surface Warfare Center have also developed a motor that includes HTSs. The field magnet for the motor consisted of two racetrack coils each supplied by ASC and IGC. The output of the motor was 167 and 112 hp at 4.2 and 27 K, respectively [8]. DOE studies have established that ≈58% of the electrical energy generated in the U.S. is used by electric motors. Because of their current-carrying capacity, HTS motors will be smaller, lighter, and more efficient than currently available motors. As a result, economic benefits associated with superconducting motors are remarkable [45].

Current Leads and Fault Current Limiters

Current leads are used to connect superconducting devices in a cryogenic environment to the outside. Because HTSs exhibit low thermal conductivity, they can minimize energy loss, reduce boiloff of the cryogen, and raise efficiency of various superconducting devices. Sumitomo Electric Industries has been using an HTS current lead that carries 2000 A for its Nb_3Ti synchrotron radiation magnets [8]. Recently, ASC and ANL developed a 16,000-A-capacity HTS current lead for Babcock & Wilcox, who anticipate using the CryoSaver™ current lead, ASC's first commercial HTS product, for an energy storage device planned for Anchorage Municipal and Power in Anchorage, Alaska [46]. In fact, the CryoSaver™ current lead is the vanguard of a series of products that ASC plans to introduce in the near future.

A sudden surge in power can render equipment or a system ineffective. Traditional approaches of using circuit breakers with higher fault current ratings to curtail such power surges have proved to be expensive and inefficient. A fault current limiter that was developed with HTSs can reduce the current to a fraction of its peak value in <8 ms, thereby mitigating the effects of short circuits or power fluctuations [8,44]. Strategic location of the fault current limiter can result in effective cost savings in improved system operation. American Superconductor Corporation, in collaboration with Lockheed Martin Corporation, SCE, LANL, and DOE, successfully demonstrated a 2.4-kV, 2.2-kA HTS fault current limiter. This is the second commercial HTS product that ASC is marketing. Plans are underway to develop, with IGC, a 15-kV, 10.6-A HTS fault current limiter [47,48].

CONCLUSION

Significant progress has been made in the development of HTSs for various applications; some of them have already made significant strides in the marketplace, whereas others are still in the developmental stages. Several parameters involved in the PIT and WIT techniques must be carefully controlled to obtain the desired properties. By improving the superconductor/silver interface, we can improve not only the strain tolerance characteristics but also the current transport properties of the HTS tapes. Long-length mono- and multifilament conductors with consistently high current transport properties have been fabricated by a carefully designed thermomechanical treatment process. An HTS magnet and a prototype transformer have been fabricated from these long conductors. An HTS magnet that contains 2400 m of tape generated a field of 3.2 T at 4.2 K and zero applied field. In addition, considerable effort is being expended to improve the characteristics of HTSs for several other applications.

ACKNOWLEDGMENTS

Work is supported by the U.S. Department of Energy (DOE), Energy Efficiency and Renewable Energy, as part of a DOE program to develop electric power technology, under Contract W-31-109-Eng-38. I thank J. G. Daley and C. E. Platt, Superconductivity Technology Program, DOE, for helpful discussions.

REFERENCES

[1]J. Bednorz and K. A. Muller, *Z. Phys. B*, **64**, 189 (1986).

[2]S. Uchida, H. Takagi, K. Kitazawa, and S. Tanaka, *Jpn. J. Appl. Phys.*, **26**, L1 (1987).

[3]M. K. Wu, J. R. Ashburn, C. J. Torng, P. H. Hor, R. J. Meng, L. Gao, Z. J. Haang, Y. Q. Wang, and C. W. Chu, *Phys. Rev. Lett.* **58**, 908 (1987).

[4]H. Maeda, Y. Tanaka, M. Fukutomi, and T. Asano, *Jpn. J. Appl. Phys.*, **27**, L209 (1988).

[5]Z. Z. Sheng and A. M. Nature, *Nature*, **332**, 55 (1988).

[6]G. J. Yurek, *Supercond. Ind.*, **9**, 12 (1996).

[7]Superonductivity for Eelctric Systems Program Plan, DOE/GO-10096-167, Department of Energy, March 1996.

[8]G. B. Lubkin, *Phys. Today*, **49**, No. 3, 48 (1996).

[9]Proceedings of Superconductivity Program for Electric Power Systems 1994 Annual Peer Review Meeting, Vol. 1, Conf.-9407153, Alexandria, VA, July 29, 1994.

[10]K. D. Ott, *Supercond. Ind.*, **7**, 10 (1994).

[11]HTS Cable Wires, News Release, American Superconductor Corporation, Westborough, MA, March 1996.

[12]G. B. Lubkin, *Phys. Today*, **48**, 20 (1995).

[13]U. Balachandran, A. N. Iyer, P. Haldar, J. G. Hoehn, Jr., and L. R. Motowidlo, in *Bismuth-Based High Temperature Superconductors*, eds., H. Maeda and K. Togano (Marcel Decker Inc., New York, 1996), p. 499.

[14]D. Dimos, P. Chaudhari, and J. Mannhart, *Phys. Rev. B*, **41**, 4038 (1990).

[15]A. S. Parikh, B. Meyer, and K. Salama, S*upercond. Sci. Technol.*, **7**, 455, (1994).

[16]S. Foner and T. P. Orlando, *Technol. Rev.*, **91**, 36 (1988).

[17]A. Bourdillon and N. X. Tan Bourdillion, in *High Temperature Superconductors* (Academic Press Inc., 1st ed., New York, NY, 1994), p. 1.

[18]P. Yam, *Sci. Amer.*, **269**, 119 (1993).

[19]M. R. Beasley, *IEEE Trans. Appl. Supercond.* **5**, 141 (1995).

[20]P. Majewski, H. Bestgen, S. Eschner, F. Aldinger, *Z. Metall.*, **86**, 563 (1995).

[21]L. Civale, A. D. Marwick, R. Wheeler IV, M. A. Kirk, W. L. Carter, G. N. Riley, Jr., and A. P. Malozemoff, *Physica C*, **208**, 137 (1993).

[22]Paul. C. W. Chu, *Sci. Amer.*, **273**, 162 (1995).

[23]K. H. Sandhage, G. N. Riley, Jr., and W. L. Carter, *J. Metals*, **43**, 21 (1991).

[24]P. Haldar and L. R. Motowidlo, *J. Met.*, **44**, 54 (1992).

[25]J. Tenbrink, M. Wilhelm, K. Heine, and H. Krauth, *IEEE Trans. Mag.*, **27**, 812 (1991).

[26]N. Schiff and E. Schempp, *Supercond. Ind.*, **7**, 14 (1994).

[27]K. Sato, T. Hikata, H. Mukai, M. Ueyama, N. Shibuta, T. Kato, T. Masuda, M. Nagata, K. Iwata, and T. Mitui, *IEEE Trans. Mag.*, **27**, 1231 (1991).

[28]Q. Li, K. Brodersen, H. A. Hjuler and T. Freltoft, *Physica C*, **217**, 360 (1993).

[29]U. Balachandran, A. N. Iyer, R. Jammy, P. Haldar, J. G. Hoehn, Jr., and M. Suenaga, in *Advances in Superconductivity VII*, eds., K. Yamafuji and T. Morishita, (Springer-Verlag, Tokyo, Japan, 1995), p. 781.

[30]D. Newman, *Supercond. Ind.*, **3**, 10 (1990).

[31]A. P. Malozemoff, Q. Li, S. Fleshler, G. N. Riley, Jr., G. Snitchler, and D. Aized, presented at the 1995 Taiwan International Conference on Superconductivity, Hualien, Taiwan, ROC, Aug. 8-11 1995.

[32]Y. Iijima, N. Tanabe, O. Kohno, Y. Ikeno, *Appl. Phys. Lett.*, **60**, 769 (1992).

[33]Proceedings of Superconductivity Program for Electric Power Systems 1995 Annual Peer Review Meeting, U.S. Department of Energy, Alexandria, VA, Aug. 1-2, 1995.

[34]M. Lelovic, P. Krishnaraj, N. G. Eror, A. N. Iyer, and U. Balachandran, *Physica C*, **242**, 246 (1995).

[35]Y. Feng and D. Larbalestier, *Interface Science*, **1**, 401 (1994).

[36]D. Larbalestier, X. Y. Cai, H. Edelman, M. B. Field, Y. Feng, J. Parrell, A. Pashitski, and A. Polyanskii, *J. Met.*, **46**, 20 (1994).

[37]L. Martini, S. Bonazzi, M. Majoros, V. Ottoboni, and S. Zannella, *IEEE Trans. Appl. Supercond.*, **3**, 961, (1993).

[38]J. Schwartz, H. Sekine, T. Asano, T. Kuroda, K. Inoue, and H. Maeda, *IEEE Trans. Mag.*, **27**, 1247 (1991).

[39]J. Ekin, in *Materials at Low Temperatures*, eds., R. P. Reed and A. F. Clark (Amer. Soc. Met., Metals Park, OH, 1st. ed., 1983), p. 494.

[40]J. P. Singh, J. Joo, N. Vasanthamohan, and R. B. Poeppel, *J. Mater. Res.*, **8**, 2458 (1993).

[41]J. Schwartz, J. K. Heuer, K. C. Goretta, R. B. Poeppel, J. Guo, and G. W. Raban, Jr., *Appl. Supercond.*, **2**, 271 (1994).

[42]U. Balachandran, A. N. Iyer, R. Jammy, P. Haldar, and M. Suenaga, *Adv. Cryogenic Eng.*, **42A**, 753 (1997).

[43]U. Balachandran, A. N. Iyer, C. A. Youngdahl, L. R. Motowidlo, J. G. Hoehn, Jr., and P. Haldar, in *Advanced Materials VI: Superconductors, Surfaces and Superlattices*, eds., H. Sakaki et al., **19A**, 385 (1994).

[44]200-hp High Temperature Superconducting Motor, News Release, Reliance Electric, Cleveland, OH, March 1996.

[45]C. Gillespie, *Supercond. Ind.*, **6**, 31 (1993).

[46]Current Leads Planned for Use by Alaskan Utility, News Release, American Superconductor Corporation, Tampere, Finland, June 1995.

[47]E. M. Leung and A. Rodriquez, *Supercond. Ind.*, **9**, 20 (1996).

[48]2.4 kV Current Limiter Prototype, News Release, American Superconductor Corporation, Westborough, MA, November 1995.

Wires and Composites

ISOTHERMAL MELT PROCESSING OF BI-2212 WIRES AND TAPES

Terry G. Holesinger
Materials Science and Technology MST-6
Los Alamos National Laboratory
MS G755
Los Alamos, NM 87545

W. Larry Hults
James L. Smith
Superconductivity and Technology Center
Los Alamos National Laboratory
MS K763
Los Alamos, NM 87545

Weiming Dai
Kenneth R. Marken
Oxford Instruments Inc.
600 Milk Street
P.O. Box 429
Carteret, NJ 07008-0429

ABSTRACT

Isothermal melt processing has been used to produce high critical currents (I_c's) and current densities (J_c's) in $Bi_2Sr_2CaCu_2O_y$ round wires and flat tapes. I_c's and J_c's of 225 A and 80 kA/cm^2 have been obtained in short length, monocore wires. However, voids and other macroscopic defects have limited performance over longer lengths. Isothermally melt processed tapes have yielded I_c and J_c values up to 341 A and 245 kA/cm^2 although performance over longer lengths is somewhat reduced. A test coil made from 0.5 m of tape yielded I_c and J_c values of 208 A and 153 kA/cm^2, respectively.

INTRODUCTION

Isothermal melt processing (IMP) is a technique by which $Bi_2Sr_2CaCu_2O_y$ (Bi-2212) composite conductors are partially melted and solidified at the same temperature [1,2]. This technique is possible due to the dependence of the melting point of Bi-2212 on the oxygen partial pressure [3]. Bi-2212 can be partially melted at temperatures above 750°C in very low oxygen partial pressures or inert atmospheres. Under these conditions, the phases present in the partial melt are $Bi_2Sr_{3-x}Ca_xO_y$ (2:3 Cu-free), CaO, and liquid [4]. By simply increasing the oxygen partial pressure after a given amount of time in the partial melt, the Bi-2212 phase is formed. IMP has been successfully applied to the processing of bulk bars, thick films, round wires, and flat tapes containing Bi-2212 [1,5,6]. In this paper, we present some of the most recent results with this technique as applied to the processing of Bi-2212 monocore round wires and multi-filament flat tape. High J_c values were obtained in both types of conductors although less variations in J_c over long lengths were obtained with the tapes.

EXPERIMENTAL

For the fabrication of the monocore round wire, powder of nominal composition $Bi_2Sr_2CaCu_2O_y$ was prepared from high-purity oxides and carbonates, calcined at 780°C, 800°C, and 825°C in air, and then splat-quenched in order to obtain a glassy precursor that is thoroughly mixed on the atomic scale. Details of this process are explained elsewhere [7]. The powder used in the current study had approximately 1.0 wt.% Al_2O_3 from reaction with the Al_2O_3 crucible during splat-quenching. Previous investigations on the effect of Al_2O_3 have shown that aluminum does not affect the superconducting properties of the Bi-2212 phase [4]. The glassy precursor was thoroughly ground and loaded into a silver or silver alloy tube with one end open. The silver alloy used was $Ag_{0.99}Cu_{0.005}Pb_{0.004}Al_{0.001}$. A silver or silver alloy plug was then inserted into the opening and welded into place. The tubes were drawn as wire to an overall diameter of 1 mm. The Bi-2212 flat tape was manufactured at Oxford Instruments. Commercial $Bi_2Sr_2CaCu_2O_y$ powder was calcined and loaded into a silver tube. The silver tube was then drawn down as wire, bundled and repacked into another silver tube, and then drawn and rolled into a 55 filament tape of overall dimensions 3.85 mm x 0.35 mm.

All wires and tapes were processed by IMP. For the round wires, 4 cm or 15 cm long lengths were heated in an ultra-high purity argon atmosphere at 10°C/min from room temperature to between 770°C and 850°C and held for 90 minutes. After this step, the gas flow was switched to a mixture of 1%O_2, 10%O_2, 27%O_2, or 100% O_2 (balances Ar). Gas flows were set so as to change the equivalent furnace-tube volume seven times per hour. After a total of 25 hours at the processing temperature, the samples were cooled at 5°C/min to room temperature. Flat tapes 15 cm in length were also processed by IMP. In this case, the partial melt was created in vacuum prior to oxidation. During the vacuum-melt step, pressures less than 1 milli-Torr were maintained in the furnace. Melting times were from 0.5 to 2.6 hours. After the melting step was complete, the furnace was backfilled quickly (< 1 min) with either 10% O_2/Ar, 20%O_2/Ar, or pure oxygen. After a total of 15 hours, the tapes were cooled at 5°C/min to room temperature. Processing temperatures for the tapes ranged from 780°C to 865°C. One 0.5 m test coil was isothermally melt processed to characterize long lengths of Bi-2212 tape.

Sample characterization was performed by quantitative optical microscopy, backscattered electron imaging in the scanning electron microscope (SEM/BEI), SQUID magnetometry, and transport measurements at 4K (helium) and 75K (nitrogen). At the altitude of Los Alamos, the boiling point of nitrogen is 75 K. Samples were prepared for microscopy and quantitative analysis by mounting the samples in cross-section and polishing using standard metallographic techniques. Cross sections of the wires were measured by quantitative optical microscopy. The critical current (I_c), critical current density (J_c), and engineering critical current density (J_e) were determined using an electric field criterion of 1.0 μV/cm. For each wire or tape, three equal lengths from the middle of the sample were measured.

RESULTS AND DISCUSSION

Figure 1 shows the dependence of J_c on the processing temperature and oxygen partial pressure used for oxidation during IMP of Bi-2212 wires. An oxidation mixture of 27% O_2/Ar consistently produced the highest I_c and J_c values. The I_c value of 130 A shown in Figure 1 corresponds to J_c and J_e values of 43 kA/cm^2 and

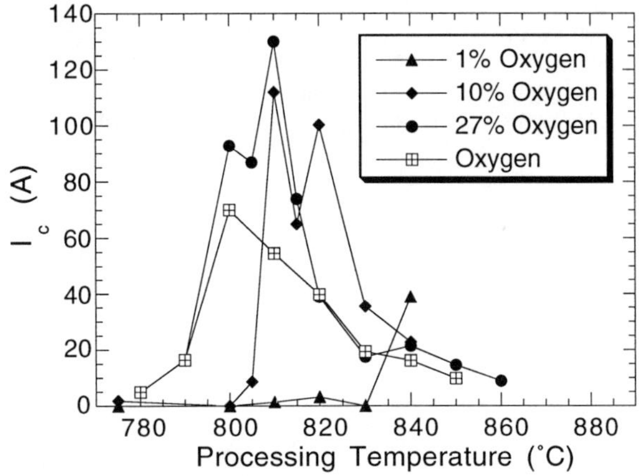

Figure 1: Plot of I_c values for Bi-2212 wires as a function of the processing temperature and oxidation atmosphere.

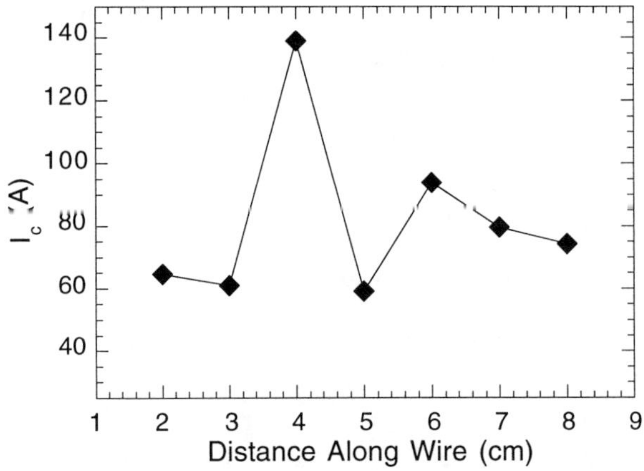

Figure 2: Variation in I_c along a long length of Bi-2212 round wire processed by IMP at 810˚C using 27%O_2/Ar as the oxidation atmosphere.

14 kA/cm^2, respectively. In all cases, the optimal processing temperature was between 800°C and 820°C. Note that these temperatures are significantly lower than those used in conventional melt processing of Bi-2212[8]. Shown in Figure 2 is the variation in I_c along a 20 cm round wire that was processed at 810°C using 27%O_2/Ar for the oxidation step of IMP. These variations in I_c along the length of long wires were common.

An attempt was made to minimize this problem with a Ag-Cu-Pb-Al alloy sheath. It was thought that an alloy would help with densification of the core during wire drawing and reduce any reactions between the silver and Bi-2212 during processing [9]. As shown in Figure 3, higher I_c values were obtained with 1 mm diameter wire containing the alloy sheath. The I_c value of 225 A translates into J_c and J_e values of 80 kA/cm^2 and 28.3 kA/cm^2, respectively at 4 K and self field. In the case of the alloy-sheathed wire, a processing temperature of 780°C and oxidation atmosphere of 10%O_2/Ar was found to be optimal. However, long lengths of wire still showed large variations in I_c. The decrease in optimal processing temperature may be related to the addition of lead to the alloy sheath.

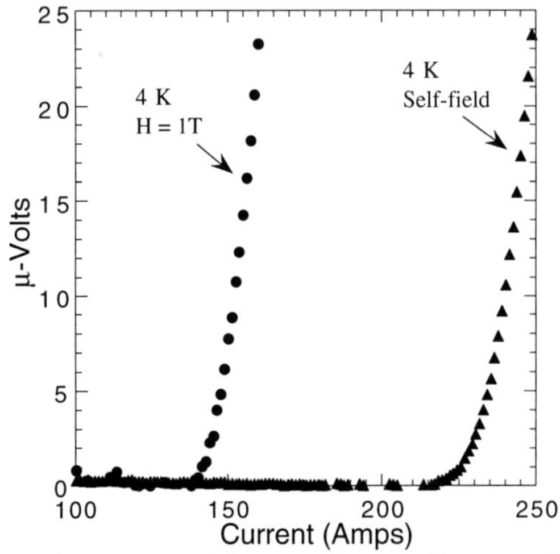

Figure 3: Current-voltage curves from a 1 mm diameter alloy-sheathed wire isothermally melt processed at 780°C using 10%O_2/Ar as the oxidizing atmosphere. The I_c values of 225 A and 140 A in self-field and an applied field of 1 T translate into J_c values of 80.4 kA/cm^2 and 50 kA/cm^2, respectively. The distance between voltage taps was 1.5 cm.

Variations in I_c were correlated with large amounts of porosity in fully processed wires. Figure 4a and b contain backscattered electron images (BEI) of as-drawn and fully processed silver-sheathed round wires. The as-drawn wire contains large cracks and significant porosity. The space represented by the cracks and porosity translates into regularly spaced voids in fully processed wires as shown in Figure 4b.

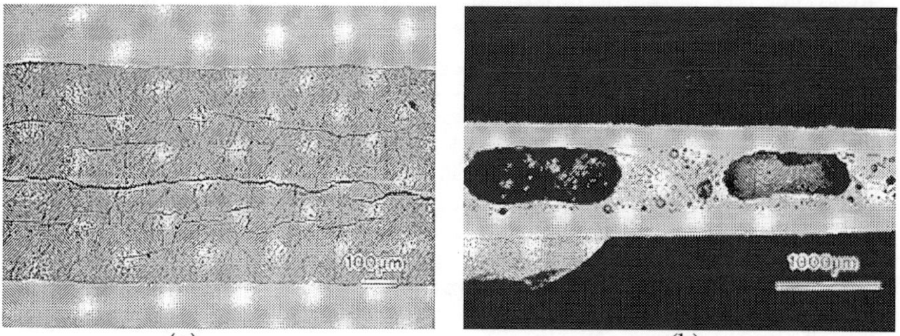

(a) (b)

Figure 4: BEI micrographs of long transverse sections of (a) as-drawn silver-sheathed wire and (b) wire isothermally melt processed at 810°C using 27% O_2/Ar.

Other microstructural features in fully processed wires are dependent upon the processing temperature. Figures 5a and b contain short transverse sections of wires processed at 780°C and 850°C, respectively. The silver-sheathed wires processed around 780°C showed signs of insufficient partial melting prior to solidification. Grain sizes were small and the porosity was uniformly distributed throughout the core. This microstructure was correlated with low J_c values. Note that with the alloy sheathed wires, the optimal processing temperature was shifted down to around 780°C. However, processing below 780°C also produced the microstructure resulting from insufficient partial melting. The wire processed at 850°C showed, in addition to the large pores, a large grain microstructure and grain-growth induced desinterring. This microstructure also correlated with low J_c values over short lengths as shown in Figure 1.

In contrast to the monocore round wire, significantly better results were obtained with multi-filament tape. This was attributed to the higher starting

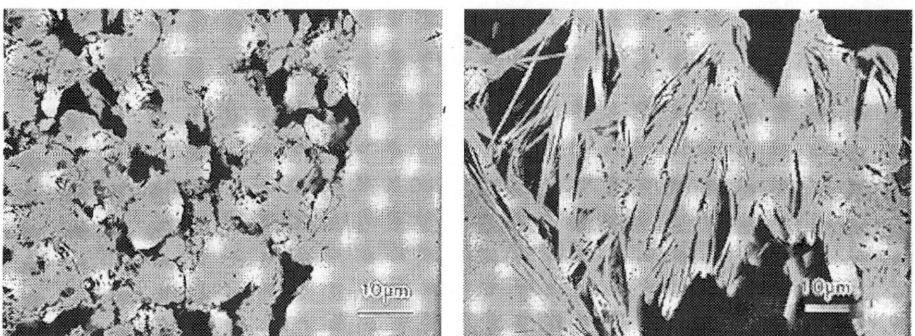

(a) (b)

Figure 5: Short transverse BEI images of silver-sheathed wires processed by IMP at (a) 780°C and (b) 850°C. The oxidation was carried out in 27% O_2/Ar.

densities obtainable in rolled tapes compared to as-drawn wire. Results presented here suggest a much larger processing region exists for obtaining high J_c tapes depending on the oxidation atmosphere, time in the partial melt, and processing temperature. For example, tapes processed at 835°C with dwell times in the partial melt of 0.8 hours and oxidation atmospheres of 100%, 20%, and 10% O_2/Ar had J_c values of 108, 212, and 177 kA/cm^2, respectively. However, processing at 820°C using melt times of 1.2 hours and oxidation atmospheres of 20% and 10% O_2/Ar yielded J_c values of 158 and 245 kA/cm^2, respectively. Although the complete processing temperature, melt time, oxidation atmosphere relationships have not been worked out, several general trends appear in the data. As the processing temperatures are decreased, the oxygen partial pressure needs to be lowered to produce high J_c tapes. In addition, higher processing temperatures require shorter partial melt times in order to maximize J_c.

The highest J_c values have been obtained at 820°C using 10% O_2/Ar as the oxidizing atmosphere. Shown in Figure 6 is a plot of the J_c values obtained in multifilament tapes as a function of time in the partial melt at 820°C. The highest I_c, J_c, and J_e values obtained across a three cm section of tape were 341A, 245 kA/cm^2, and 25.6 kA/cm^2, respectively. The window for obtaining high J_c tapes is relatively wide. Melt times of 1 to 1.5 hours were found to give high J_c values. The standard deviations associated with the J_c values do not appear to correlate with the melt times or porosity. Hence it was assumed that the variations in I_c were due to local inhomogenities in the starting tape. To test IMP on longer lengths of tape, a 0.5 m test coil was processed at 820°C using 10% O_2/Ar for oxidation. The time spent in the partial melt was 1.3 hours. The overall I_c and J_c of the coil was 213 A and 153 kA/cm^2. The coil values are consistent with the variations observed in the short lengths of tape.

Secondary phases were present in all fully processed tapes to various degrees.

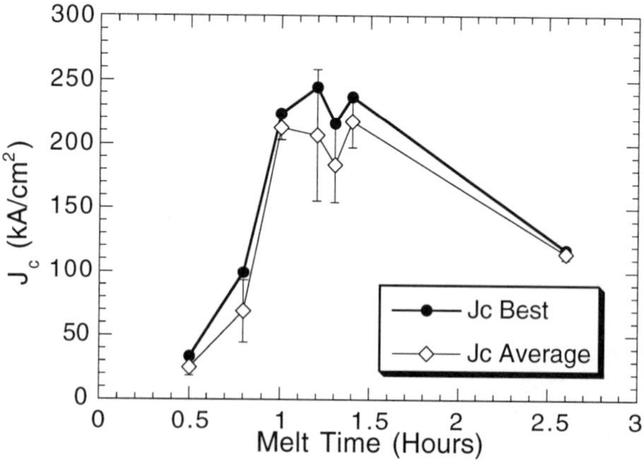

Figure 6: J_c values as a function of time spent in the partial melt for Bi-2212 tapes processed by IMP at 820°C using 10% O_2/Ar for oxidation. The best values are from 3 cm sections of long tapes. The average J_c and corresponding standard deviation represent all measurements from three sections of two or more tapes.

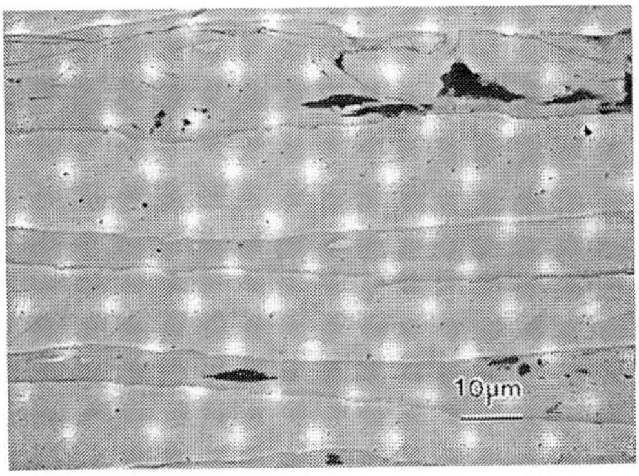

Figure 7: Short transverse SEM/BEI image of a Bi-2212 processed by IMP at 820°C using 10% O_2/Ar for oxidation. Time spent in the partial melt was 1.3 hours. A region containing large secondary phases is evident at the top of the micrograph.

Unlike the round wire, the tapes appeared to be dense and free of large pores. Shown in Figure 7 is a short transverse SEM/BEI image of a tape processed at 820°C for 1.3 hours and oxidized in 10% O_2/Ar. The J_c value for this tape was 216 kA/cm^2 at 4 K and self-field. For the most part, the filament uniformity and phase purity were quite good although areas could be found where large secondary phases were concentrated. This random arrangement of the pockets of secondary phases is presumed to result from inhomogenities in the starting tape. Further work is in progress to verify and minimize this problem.

The samples with low J_c values in Figure 6 were found to have microstructures that resulted from either spending too little or too much time in the partial melt. Similar microstructures were encountered with the round wires. Figures 8a and b contain SEM/BEI images of tapes processed at 820°C where the times spent in the

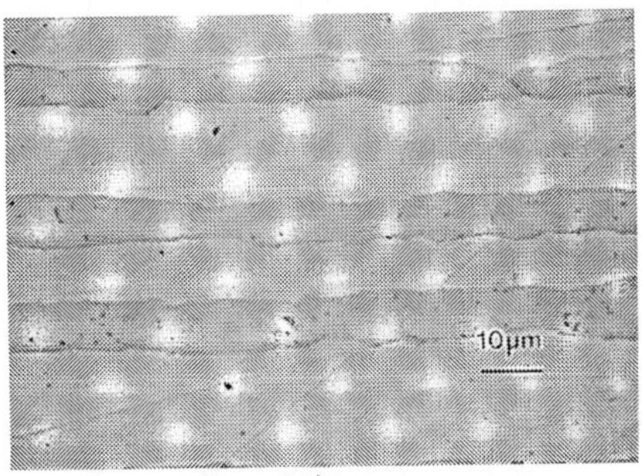

(a)

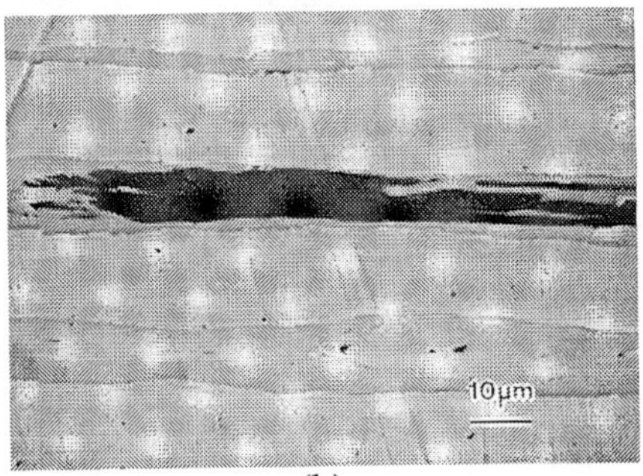

(b)

Figure 8: Long transverse SEM/BEI images of Bi-2212 tapes processed by IMP at 820°C using 10% O_2/Ar for oxidation. The times spent in the partial melt were (a) 0.5 hours and (b) 2.6 hours.

partial melt were 0.5 and 2.6 hours, respectively. In Figure 8a, the tape that was partially melted for only 0.5 hours shows very good phase purity. This tape contained the fewest secondary phases of all of the tapes processed at 820°C. Unfortunately, the I_c values were very low. With the short times in the partial melt, an insufficient amount of liquid is formed and the subsequent solidification step does not result in a well aligned, connected microstructure such as that shown in Figure 7. For the tape that spent 2.6 hours in the partial melt, a uniform distribution of large alkaline-earth cuprate needles were found throughout the sample. These large secondary phases were often found to span the entire thickness of the filaments as shown in Figure 8b.

One of the advantages of the IMP technique for Bi-2212 processing is that the filament uniformity can be maintained throughout the process. Figure 7 shows a high-J_c tape where the Ag-Bi-2212 interface is smooth and well-defined with no grain growth into the silver. With conventional melt processing, maintaining filament uniformity is often a problem [10]. Peak temperatures in conventional melt processing can reach as high as 885°C to 920°C. At these temperatures, the silver sheath softens significantly and grain growth of Bi-2212 into the sheath material is often observed. Grain growth into the silver sheath was observed with IMP when processing temperatures of 850°C or greater were used. Shown in Figure 9 is a SEM/BEI image of a Bi-2212 tape processed by IMP at 865°C using oxygen. Time spent in the partial melt was 0.9 hours and the J_c value of the tape was 69.4 kA/cm^2. A common microstructural feature found in these latter tapes is the presence of grain colonies cutting across the filaments as shown in Figure 9. Filament cutting by misaligned colonies was not observed in tapes processed by IMP at temperatures below 850°C.

SUMMARY

Isothermal melt processing has been applied to the processing of Bi-2212 monocore round wires and multifilament tapes. IMP produced high J_c values up to

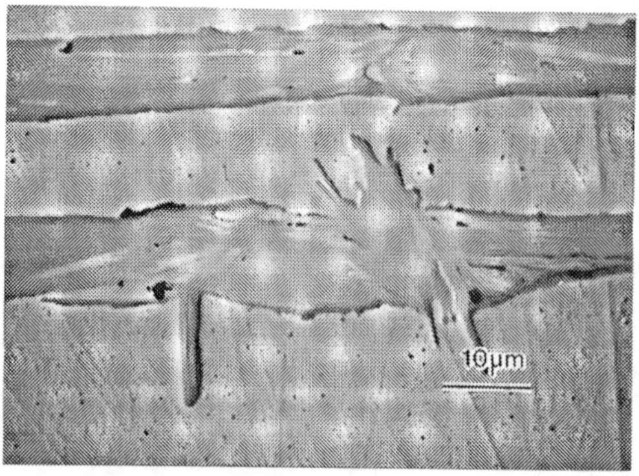

Figure 9: A long transverse SEM/BEI image of a Bi-2212 tape processed by IMP at 865°C using oxygen. Time spent in the melt was 0.9 hours. The misaligned colony of grains seen cutting across the filament was a common feature in tapes processed above 850°C.

80 kA/cm^2 at 4 K and self field in short lengths of round wires. However, large amounts of residual porosity caused significant variations in I_c over long lengths. Such problems with the multifilamentary tapes were not encountered because of the higher starting density obtainable in rolled tapes. J_c values in tapes processed by IMP were found to be dependent upon the processing temperature, time spent in the partial melt, and the oxidation atmosphere. The highest J_c values were obtained in tapes processed by IMP at 820°C. I_c and J_c values of 341 A and 245 kA/cm^2 were obtained over short sections. A 0.5 m test coil processed under similar conditions was found to have I_c and J_c values of 213 A and 153 kA/cm^2, respectively. Variations in I_c in the tapes were attributable to localized inhomogenities in the starting material. It was found that filament uniformity was maintained in tapes processed by IMP when the processing temperature was kept below 850°C. Processing temperatures of 850°C or greater tended to produce misaligned colonies that cut across individual filaments.

ACKNOWLEDGMENTS

This work was performed under the auspices of the United States Department of Energy, Office of Energy Management and Renewable Energy, as part of a DOE program to develop electric power technology, under contract W-7405-ENG-36. The authors would like to acknowledge the help of John Bingert and Barry Bingham in the fabrication of Bi-2212 round wires.

REFERENCES

[1]T.G. Holesinger, D.J. Miller, H.K. Viswanathan, K.W. Dennis, L.S. Chumbley, P.W. Winandy, and A.C. Youngdahl, "Directional Isothermal Growth of Highly Textured $Bi_2Sr_2CaCu_2O_y$," *Applied Physics Letters,* **63** [7] 982-984 (1993).

[2]J. MacManus-Driscoll, Pin-Chin Wang, J.C. Bravman, and R.B. Beyers, "Phase Equilibria and Melt Processing of $Bi_2Sr_2CaCu_2O_{8+x}$ Tapes at Reduced Oxygen Partial Pressures,"*Applied Physics Letters,* **65** [22] 2872-75 (1994).

[3]P. Strobel, W. Korczak, and T. Fourneir, "A Thermal-Analysis Study of the System Bi-Sr-Ca-Cu-O at Variable Oxygen-Pressure,"*Physica C,* **161** [2] 167-174 (1989).

[4]T.G. Holesinger, "Al_2O_3 Additions for Isothermal Melt Processing of $Bi_2Sr_2CaCu_2O_y$," *Journal of Materials Research,* **11** [9] 2135-41 (1996).

[5] T.G. Holesinger, D.S. Phillips, J.Y. Coulter, J.O. Willis, and D.E. Peterson, "Isothermal Melt Processing of Bi-2212 Thick Films," *Physica C,* **243** 93-102 (1995).

[6]T.G. Holesinger, J.M Johnson, J.Y. Coulter, H. Safar, D.S. Phillips, J.F. Bingert, B.L. Bingham, M.P. Maley, J.L. Smith, and D.E. Peterson, "Isothermal Melt Processing of $Bi_2Sr_2CaCu_2O_y$ Round Wire," *Physica C,* **253** 182-190 (1995).

[7]T.G. Holesinger, D.J. Miller, and L.S. Chumbley, "Crystallization of Bi-Sr-Ca-Cu-O Glasses in Oxygen," *Journal of Materials Research,* **7** [7] 1658-1671 (1992).

[8] Y. Feng, K.E. Hautanen, Y.E. High, D.C. Larbalestier, R. Ray II, E.E. Hellstrom, and S.E. Babcock, "Microstructural Analysis of High Critical Current Density Ag-Clad Bi-Sr-Ca-Cu-O (2:2:1:2) Tapes," *Physica C* **192** 293-305 (1992).

[9] B.N. Hubert, R. Zhou, T.G. Holesinger, W.L. Hults, A. Lacerda, A.S. Murray, R.D. Ray II, C.M. Buford, L.G. Phillips, A. Kebede, and J.L. Smith, "Silver Alloys for High-Temperature Superconducting Wire," *Journal of Electronic Materials*, **24** [12] 1869-1872 (1995).

[10] L.R. Motowidlo, G. Galinski, G. Ozeryansky, W. Zhang, and E.E. Hellstrom, "Dependence of Critical Current Density on Filament Diameter in Round Multifilament Ag-Sheathed $Bi_2Sr_2CaCu_2O_y$ Wires Processed in O_2," *Applied Physics Letters*, **65** [21] 2731-2733 (1994).

COATED-WIRE-IN-TUBE PROCESSING OF BISMUTH–2223 SUPERCONDUCTORS

S. E. Dorris, N. Ashcom, T. Truchan, and N. Vasanthamohan
Argonne National Laboratory, Argonne, IL 60439–4838

D. A. Burlone
Fiber Products Division, BASF Corporation, Enka, North Carolina 28728

L. D. Woolf
General Atomics, San Diego, CA 92186-9784

ABSTRACT

A coated-wire-in-tube (CWIT) process greatly increases the silver/superconductor interface area in silver-sheathed Bi-2223 superconductors. When the performance of CWIT samples is compared to that of conventional monofilaments made with the same powder, critical current density increases significantly with increased silver/superconductor interface area. Benefits of increasing the silver/superconductor interface area are realized only when there is good continuity of the coated wires, which can be preserved by proper mechanical processing.

INTRODUCTION

Abundant evidence [1-5] suggests that a large fraction of the current in silver-sheathed Bi-2223 conductors is carried by dense, well-aligned material at the silver/superconductor interface. In an effort to improve the performance of Bi-2223 conductors, a coated-wire-in-tube (CWIT) process has been developed to greatly increase the silver/superconductor interface area. In CWIT processing, fine silver wires (0.003-0.020 in. diameter) are coated with Bi-2223 precursor powder (coating thickness ≈25-100 μm) by either a slurry method [6,7] or electrophoresis [8]. In the slurry method, the Bi-2223 coating is tough and adherent due to its relatively high concentration of organics (≈25 vol.%), but the removal of organics may be difficult. Coatings made by electrophoresis are not as adherent as those made by the slurry method, but they contain much lower concentrations of organics and therefore removal of the organics is not as difficult. With either method, 100-600 coated wires are bundled and loaded into a silver tube, then drawn and rolled into a tape in a manner similar to the powder-in-tube (PIT) method.

In addition to increasing the silver/superconductor interface area, CWIT processing has several other possible advantages over alternative methods for fabricating silver-sheathed conductors. The matrix of CWIT conductors consists of superconductor, whereas silver forms the matrix of multifilament conductors, another type of conductor with an increased silver/superconductor interface area. As a result, percolation through CWIT conductors may be better than that in multifilament conductors, in which small-scale disruptions such as second-phase particles or cracks may force transport through silver. In addition, the superconductor layer thickness can be reduced in CWIT conductors, and the silver/superconductor interface area increased, by decreasing the initial coating thickness. This means that thinner superconductor layers can be made without additional mechanical deformation. By comparison, the filaments of a multifilament conductor are made finer by subjecting them to additional mechanical deformation, which can roughen the interface and cause misalignment of grains in the interface layer. Other expected benefits of CWIT processing include good strain tolerance due to the distribution of silver on a fine scale, and the ease of adapting it to the fabrication of long-length-conductors, perhaps even by a continuous process.

EXPERIMENTAL

Bundles of coated wires were prepared, some from wires made by slurry coating, others from wires coated by electrophoresis. The bundles, each containing 125-150 coated wires, were loaded into silver tubes (4.35 mm I.D. x 6.35 mm O.D. x 250 mm); the organics were then removed from the coating by heating (5°C/h up to 720°C, then hold for 24 h) under a partial vacuum (2-3 torr of flowing oxygen). The partial vacuum facilitates removal of gaseous products and prevents decomposition of the superconductor during the process. After the tubes with the slurry-coated wires were heated, both ends of all tubes were swaged so that the coated wires were gripped during mechanical deformation. In addition to the CWIT samples, a monofilament was prepared by the conventional PIT process; this sample served as a reference for the CWIT samples. The CWIT samples and the monofilament were all prepared from the same powder, $Bi_{1.8}Pb_{0.4}Sr_{2.0}Ca_{2.0}Cu_{3.0}O_{10}$ prepared by Argonne's two-powder process [9].

Mechanical processing consisted of drawing the tubes from 0.635 cm to 0.2 cm at a 10% reduction in diameter per pass, which corresponds to ≈19% reduction in area per pass and ≈23% elongation. The resulting wires were then rolled to the final tape thickness at 10-12% reduction in thickness per pass, with annealing for 5 min at 400°C after each drawing and rolling reduction. The final tapes were ≈250 μm thick and ≈4 mm wide. For later reference, this deformation sequence is called the "harsh" process. To reduce the failure of silver wires during mechanical deformation, a "gentle" process was developed for processing subsequent CWIT samples. In this process, wires were drawn to 0.2 cm diameter at 3% reduction in diameter per pass (6% reduction in area and ≈6% elongation per pass), then rolled to ≈250 μm at <10% elongation per pass. All samples made by the gentle process were prepared from wires by electrophoresis.

Samples (≈3 cm long) were cut from the tapes and heated for 50 h at 818-823°C in 8% O_2/balance N_2, after which the tapes were uniaxially cold-pressed at ≈2 GPa. Following pressing, the tapes were annealed for an additional 100 h at 818-823°C, pressed one more time, then heated for 100 h at 818-823°C. Critical current, i_c, was measured by the four–probe method at 77 K in zero applied field with a 1 μV/cm criterion. Superconductor area was measured by digital image analysis of transverse sections. To calculate critical current density, J_c, of Bi-2223, i_c was divided by the superconductor area determined by digital image analysis. The engineering critical current density, J_e, was calculated by dividing i_c by the cross-sectional area of the tape (sheath + core).

RESULTS

Figure 1 shows i_c versus total heat treatment time for tapes prepared by the harsh mechanical deformation sequence. Each data point represents the average of measurements from three individual tapes. For each tape thickness, monofilaments had the highest i_c and CWIT samples made by slurry-coating had the lowest i_c. For each tape thickness, the CWIT sample made by electrophoresis had an i_c that was ~50% higher than the i_c of slurry-coated CWIT samples. While there is no conclusive evidence that removal of organics degraded the performance of slurry-coated samples, the microstructures of those samples were similar to those of previous PIT samples that had high carbon contents, exhibiting larger and more abundant second phases in addition to significant amounts of Bi-2212. This suggests that organic removal may have been a factor in decreasing the J_c values of slurry-coated CWIT samples relative to those of the electrophoresis samples. For this reason, subsequent study of the CWIT process focused on wires coated by electrophoresis.

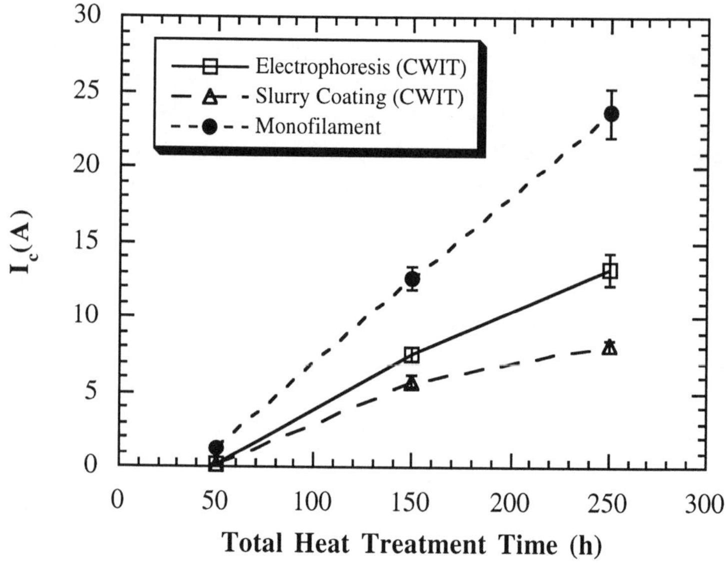

Fig. 1. Critical current, i_c, vs. total heat treatment time for samples processed by "harsh" mechanical deformation sequence. Tape thickness before heat treatment was 0.010 in. Each data point represents average of measurements from three individual tapes.

The i_c of the CWIT electrophoresis sample was ≈50% of that of the i_c of the monofilament sample (Fig. 1). Digital image analysis of transverse sections showed that the superconductor area was ≈50% smaller in the CWIT sample than in the monofilament, so J_c of the superconductor was approximately the same in the CWIT and monofilament tapes, indicating that the silver wires had not enhanced superconductor performance. Examination of longitudinal sections revealed a thin (1-2 μm), dense layer of Bi-2223 at the interface with the silver wires, but the wires were broken in many places (Fig. 2) and were separated from one another by many grain lengths. This suggested that current was forced through regions with randomly oriented grains in order to pass from one well-aligned region to the next. In turn, this led to speculation that such a microstructure may reduce or eliminate the benefit of the silver wires and that the silver wires would be beneficial only if they were continuous, i.e., if failure of the silver wires could be prevented.

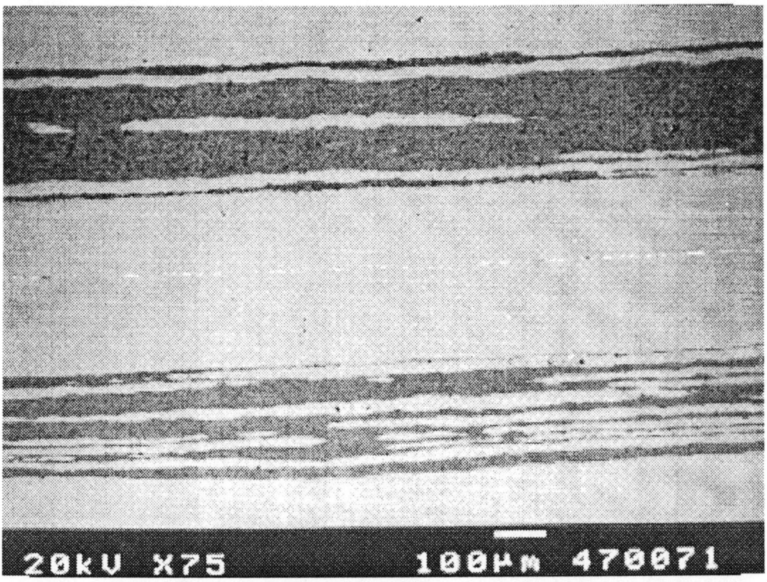

Fig. 2. Secondary electron image of two CWIT samples (longitudinal sections) prepared by "harsh" mechanical processing. Dark gray regions are Bi-2223; light gray regions are silver sheath along edges, and silver wires in core (which are broken in many places).

Figure 3 suggests that the silver wires inside CWIT conductors may have broken because of excessive deformation between annealing and/or inadequate annealing. Figure 3a plots ductility (represented by percent elongation per 2-in. sample) as a function of degree of cold-working (represented by percent reduction in area of 0.091-in.-diameter fine silver wire) [10]. For wire that has undergone a 19% reduction in area, an elongation of ≈20% is expected, and this is approximately the elongation experienced by CWIT wires during a single reduction. This suggests that the silver wires are in a critical regime after the first reduction and that subsequent elongation must be carefully limited to prevent failure of the wires. Figure 3b shows ductility versus annealing temperature (annealing time = 0.5 h) for silver wire subjected to 50% cold-working [10]. Annealing for 0.5 h at 400°C clearly imparts sufficient ductility to the silver wires to withstand 20% elongation, but annealing for 5 minutes (as done in "harsh" processing) may be inadequate, especially considering that the wires are embedded in a porous ceramic matrix that impedes thermal conduction.

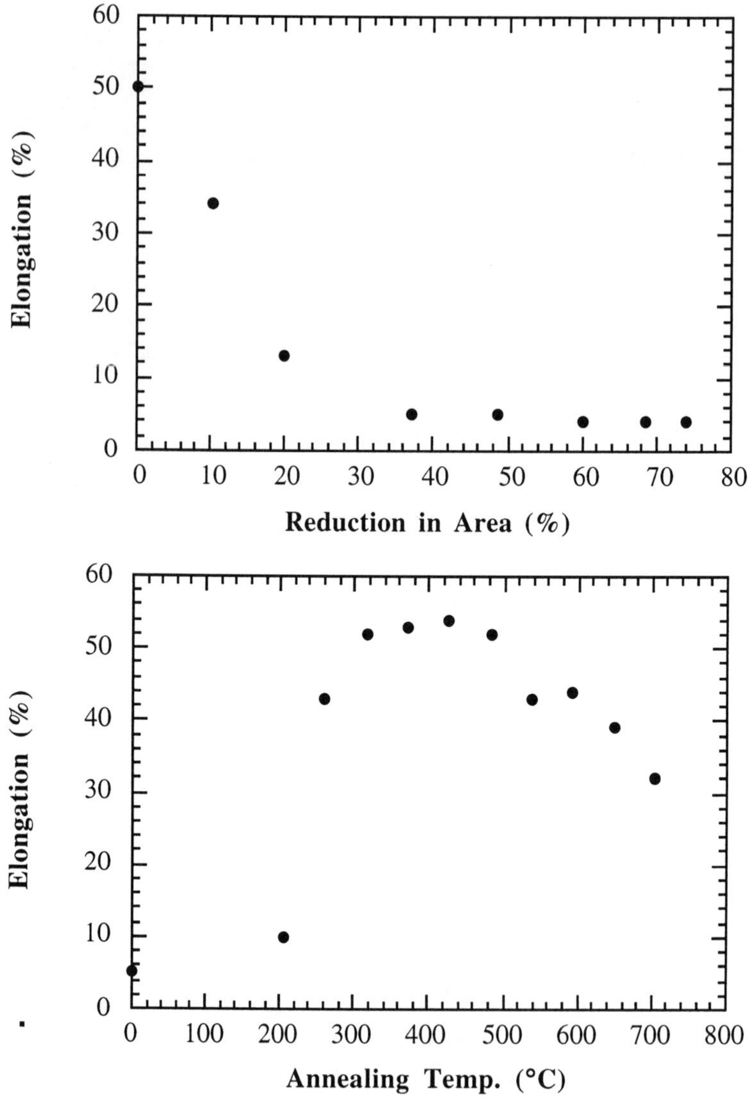

Fig. 3. Ductility of 0.091-in.-diameter fine silver wire, given as percent elongation for a 2-in. specimen, vs. (a) degree of cold-working, given as reduction in area (%), and (b) annealing temperature, where a sample subjected to 50% cold-working is annealed for 0.5 h at various annealing temperatures. Data taken from Ref. (10).

To prevent failure of the silver wires, the gentle mechanical deformation process was developed in which the reduction per pass was reduced and the annealing time was increased. Details of the gentle process are described in the experimental section. As seen in Fig. 4, the gentle deformation sequence nearly eliminated the failure of silver wires in CWIT conductors. Figure 5 compares transverse views of typical CWIT and monofilament conductors. Measuring the total silver/superconductor interface (measured as a length in the two-dimensional transverse view) and dividing by the area of superconductor showed that the interface per unit area of superconductor increased from $\approx 40\ mm^{-1}$ in the monofilament to $\approx 450\ mm^{-1}$ in the CWIT conductor. The core areas (including superconductor and silver wires) are approximately the same for the two conductors, but the superconductor area is ≈50% smaller in the CWIT conductor because silver accounts for ≈50% of the CWIT core. Despite the smaller superconductor area, i_c was ≈50% higher in the CWIT conductor, showing that superconductor performance had been enhanced.

J_c is plotted in Fig. 6 as a function of initial tape thickness (i.e., before heat treatment). J_c (Bi-2223) was ≈150% higher in the CWIT sample than in the monofilament; J_e was ≈50% higher for the CWIT conductor. These results support the idea that material at the silver/superconductor interface has improved properties relative to material at the core interior, and demonstrate that J_c and J_e of tape conductors can be enhanced by increasing the silver/superconductor interface area. In addition, because the CWIT conductor had a higher J_c than the monofilament when the silver wires remained continuous, but a lower J_c when the wires were broken, the silver wires are required to have good continuity in order for the CWIT approach to be beneficial.

CONCLUSIONS

A coated-wire-in-tube (CWIT) method greatly increases the interface area between silver and superconductor in Bi-2223 silver-sheathed superconductors. Using this CWIT process, we increased the silver/superconductor interface area by more than one order of magnitude. As a result, critical current density (J_c) was ≈150% higher in the CWIT sample than in a monofilament made with the same powder, and engineering critical current density was ≈50% higher in the CWIT conductor. These results clearly indicate that the performance of silver-sheathed Bi-2223 superconductors can be enhanced by maximizing the interface area between silver and superconductor. We also demonstrated that the mechanical deformation sequence is important to the success of the CWIT process. "Harsh" mechanical processing caused many breaks in the silver wires, and consequently the

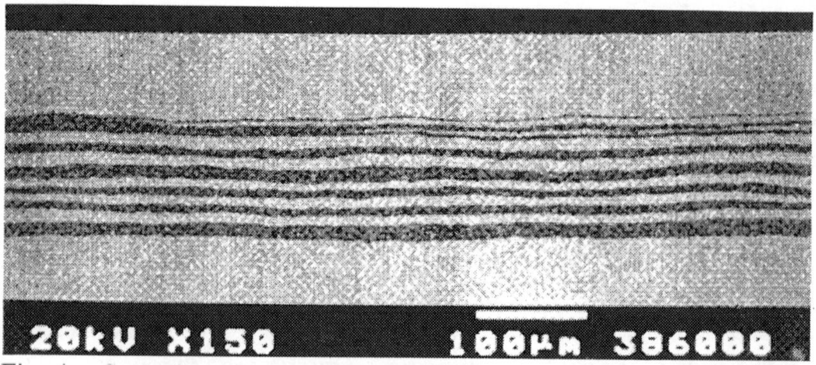

Fig. 4. Secondary electron image of CWIT sample (longitudinal section) prepared by "gentle" mechanical processing. Dark gray regions are Bi-2223; light gray regions along edges are silver sheath, and light gray regions in core are silver wires.

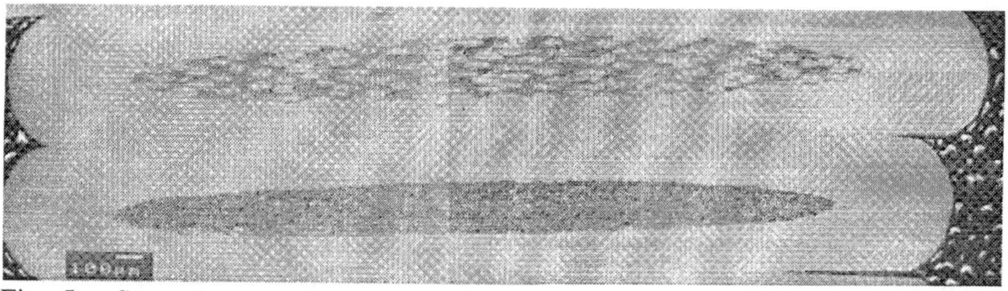

Fig. 5. Secondary electron images (transverse sections) of monofilament (bottom) and CWIT (top) tapes prepared by "gentle" mechanical processing. Dark gray regions are Bi-2223; light gray regions along the edges are silver sheath; light gray regions in the CWIT core are silver wires.

well-aligned superconductor at the interface was not continuous. As a result, incorporation of the silver wires did not enhance the performance of the superconductor. By contrast, "gentle" mechanical processing preserved the continuity of the silver wires, and incorporation of the silver wires increased the superconductor J_c.

ACKNOWLEDGMENTS

Work was supported by the U.S. Department of Energy, Energy Efficiency and Renewable Energy, as part of a program to develop electric power technology, under Contract W–31–109–Eng–38.

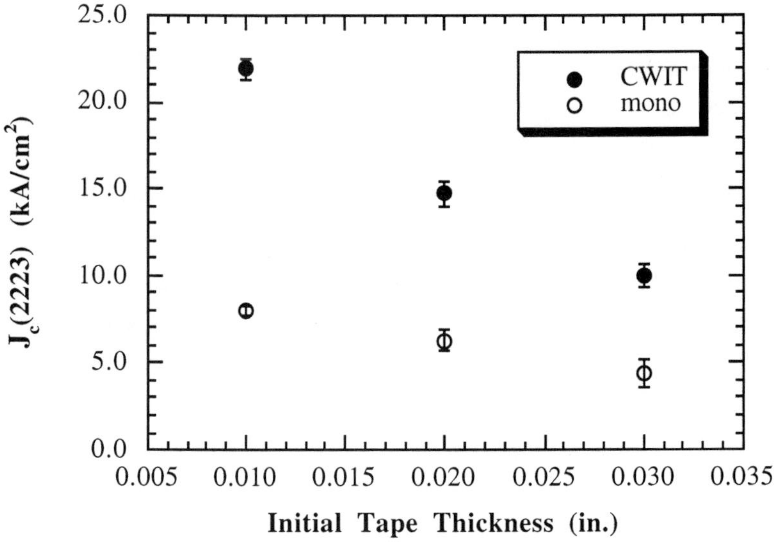

Fig. 6. Critical current density, J_c, vs. initial tape thickness (i.e., before heat treatment) of monofilament and CWIT tapes made by "gentle" mechanical processing.

REFERENCES

[1]A. E. Pashitski, A. Polyanskii, A. Gurevich, J. A. Parrell, and D. C. Larbalestier, Physica C, 246 (1995), 133.

[2]G. Grasso, B. Hensel, A. Jeremie, and R. Flukiger, Physica C, 241 (1995), 45.

[3]H. K. Liu, R. K. Wang, and S. X. Dou, Physica C, 229 (1994) 39.

[4]U. Welp, D. O. Gunter, G. W. Crabtree, W. Zhong, U. Balachandran, P. Haldar, R. S. Sokolowski, V. K. Vlasko-Vlasov, and I. Nikitenko, Nature, 376 (1995) 44.

[5]M. Lelovic, P. Krishnaraj, N. G. Eror, and U. Balachandran, Physica C, 242 (1995) 246.

[6]Practical Superconductor Development for Electrical Power Applications, Argonne National Laboratory, Annual Technical Status Report, 1995.

[7]U.S. Patent Application 08/504,330. Filed July 19, 1995.

[8]L. D. Woolf, T. L. Figueroa, R. A. Olstad, F. E. Elsner, and T. Ohkawa, J. Electronic Materials, 24(12), 1797 (1995).

[9]S. E. Dorris, B. C. Prorok, M. T. Lanagan, S. Sinha, and R. B. Poeppel, Physica C, 212 (1993) 66.

[10]Silver, Economics, Metallurgy and Use, ed. by A. D. Butts and C. D. Coxe, D. Van Nostrand Co. Inc., Princeton, NJ (1967).

STRESS/STRAIN PERFORMANCE OF SHEATHED BISMUTH-2223 SUPERCONDUCTORS

N. Vasanthamohan, J. P. Singh, G. Khanna, and M. T. Lanagan
Energy Technology Division
Argonne National Laboratory
Argonne, Illinois 60439

ABSTRACT

Besides possessing high current transport properties, superconductors employed in several applications must also be capable of retaining those properties when subjected to stresses and strains. This paper explores the potential for Consil, an alloy containing Ag, 0.25 wt.% Ni and 0.25 wt.% Mg, to serve as a potential sheathing material in the processing of BSCCO (2223) superconductors. Strain tolerance measurements at 77 K and 0 Tesla indicate superior performance by Consil-sheathed tapes over tapes sheathed in commercially pure Ag. While the pure Ag undergoes an abrupt decrease in critical current (I_c) beyond a certain strain level (0.2% strain), Consil-sheathed tapes show a more gradual decrease. However, despite their superior performance under strain, the Consil-sheathed tapes did not yield critical current density (J_c) values that are normally obtained in Ag-sheathed tapes. The reason for this behavior has been attributed to poor phase conversion of the superconducting core inside the Consil sheath, even after extended sintering times (250 h).

INTRODUCTION

Because high-temperature superconductors are brittle, they must be processed together with metals as a composite so that they can be fabricated with established metal-working techniques. Of the many metals that have been evaluated, silver is the most promising in terms of processing and chemical compatibility. It serves as a good sheathing material by containing the vapor phases, protecting the core from contact with the atmosphere, and providing structural stability. Moreover, Ag does not degrade the critical transition temperature or disrupt the stability of the superconducting phases. Instead, there is evidence that most of the supercurrent is transported along the area adjacent to the Ag sheath [1].

Despite the above advantages, Ag is soft and ductile and does not provide the necessary strength for practical applications. To make conductors robust enough to handle stresses and strains during handling and operation, multifilamentary [2,3] conductors and conductors with various alloy sheaths have been developed [4-8]. The obvious choice among candidate materials to replace Ag as a sheathing material would be alloys of Ag, preferably those that are available commercially. Consil, containing Ag, 0.25 wt.% Mg, and 0.25 wt.% Ni is one such alloy that is available commercially and can be hardened by a simple internal oxidation treatment.

Because the tapes are sintered in an oxidizing atmosphere, a separate oxidation treatment is not necessary to harden the sheath. Of the numerous factors that may influence the performance of BSCCO(2223) superconductors, phase purity, grain alignment, core density, and intergrain connectivity are known to play critical roles in determining the superconducting properties of the tape [9]. Studies have shown that Ag aids in improving grain alignment, promotes liquid-phase formation that enhances Bi-2223 formation kinetics and increases core density [10-11]. In an attempt to find a suitable stronger sheathing material, one should also ensure that the above-mentioned advantages are still offered by the alternative sheath. In this study, Consil and Ag-sheathed powder-in-tube tapes have been processed in parallel to determine the compatibility of Consil and its influence on microstructure and superconducting properties. The yield strength and ductility of Ag- and Consil-sheathed tapes were measured at room temperature (RT) and at 77 K. The tensile strain tolerance was measured at 77 K on tapes that were sintered for 50 h.

EXPERIMENTAL

The nominal composition of the superconductor used in this study is $Bi_{1.8}Pb_{0.4}Sr_2Ca_2Cu_3O_{10}$. The process follows a two-powder approach, wherein $Bi_{1.8}Pb_{0.4}Sr_2Ca_1Cu_2O_8$ and $CaCuO_2$ are reactively sintered within the sheath to give $Bi_{1.8}Pb_{0.4}Sr_2Ca_2Cu_3O_{10}$. The $Bi_{1.8}Pb_{0.4}Sr_2Ca_1Cu_2O_8$ and $CaCuO_2$ were prepared by calcining appropriate proportions of the elemental oxides and carbonates under reduced oxygen pressure. The stoichiometric mixture is packed in two different tubes (Ag and Consil) to ≈30% theoretical density, and vacuum-annealed at 300°C to remove moisture that may have been introduced during packing. Vacuum annealing at higher temperatures led to undesirable hardening and loss of ductility of the Consil tube, resulting in poor ductility and hence, difficult mechanical processing. The tubes were then sealed and mechanically processed into a tape by swaging, drawing, and rolling. The tapes were subsequently sintered in an atmosphere containing 8% O_2 (balance N_2) at 815°C for 150 h with an intermediate pressing (2 GPa for 5 s) after the first 50 h of sintering.

The strength and ductility of the tapes were measured by a simple tensile test at RT and at 77 K (in liquid N_2). Critical current, I_c was measured by the standard four-probe technique with a 1 μV/cm criterion. Strain tolerance in tension was measured in-situ by mounting the tapes between fixtures on an Instron machine with a liquid N_2 reservoir mounted on top of the moving crosshead. Critical current measurements were taken by temporarily holding the crosshead at regular strain intervals. The samples were analyzed for their composition and microstructure with a scanning electron microscope (SEM) equipped with a NORAN energy-dispersive spectrometer. X-ray diffraction (XRD) analysis was carried out with a Scintag 2000 diffractometer on exposed tape cores to monitor the Bi-2223 phase formation.

RESULTS AND DISCUSSION

Heat Treatment of Tapes

Ag-sheathed tapes were sintered at 815°C in a 8% O_2 atmosphere for 150 h with one intermediate pressing after the first 50 h of sintering. Because the Consil-sheathed tapes did not attain phase purity after 150 h of sintering, they were sintered for an additional 100 h. A plot of critical current density, I_c, versus sintering time for Consil-, and Ag-sheathed Bi-2223 tapes is shown in Fig. 1. The plot shows a significant difference in I_c between the Ag and Consil sheaths. This is attributed to the influence of the sheathing material on 2223 formation and microstructure. Figure 2 shows the XRD patterns of Ag- and Consil-sheathed tapes sintered for 150 and 250 h, respectively. The patterns indicate that the conversion to the 2223 phase is incomplete in the Consil-sheathed tape. The 2223 content was estimated from the ratio of the corresponding (0 0 l) peak area of the 2223 phase to that of the 2212 phase.

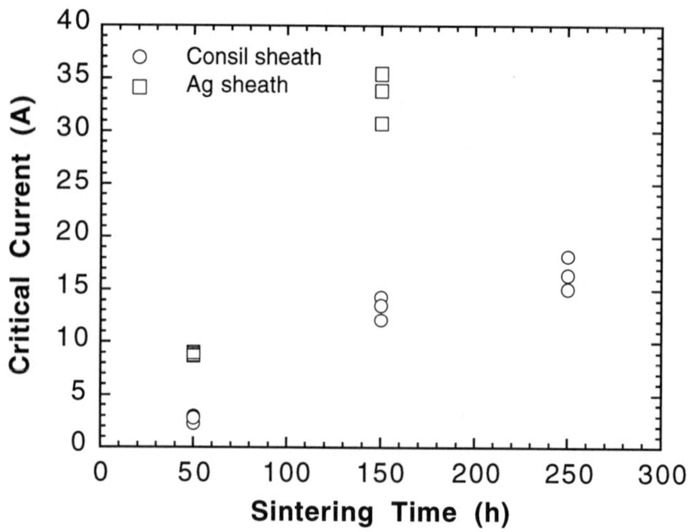

Fig. 1. I_c as a function of time for Ag- and Consil-sheathed tapes. Ag-sheathed tapes were not sintered longer than 150 h because additional sintering times did not lead to a significant increase in I_c.

The XRD patterns show that the amount of 2223 formed in the Consil tapes did not exceed 75% even after 250 h of sintering, whereas the Ag-sheathed tapes attained maximum phase purity (>95% 2223) within 150 h of sintering. The XRD pattern of Consil also shows extensive alkaline-earth phases, indicating possible decomposition of the Bi-2223 phase.

Figure 3 (a) and (b) shows backscattered electron micrographs of Ag- and Consil-sheathed tapes after 150 h of sintering. The micrographs confirm the poor phase purity that was indicated by the XRD patterns (Fig. 2). The microstructures also show that the Ag-sheathed tape has long aligned Bi-2223 grains, compared to the Consil-sheathed tape which appears to contain copious amounts of secondary phases. The individual Bi-2223 grains in the Consil-sheathed tapes appear to be far less textured than those in the Ag-sheathed tapes. The dense microstructure serves as evidence for liquid-phase formation that subsequently led to decomposition of the high-T_c phase. The poor phase purity and resulting decreased I_c values can be attributed to one or more of the following:

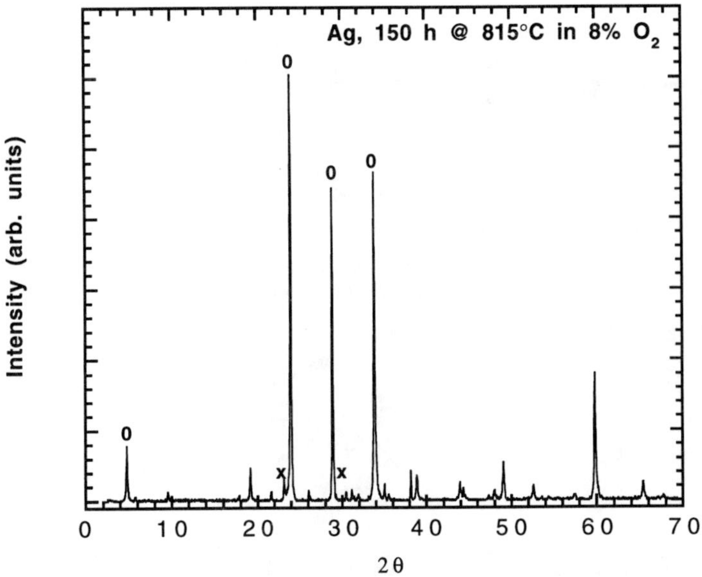

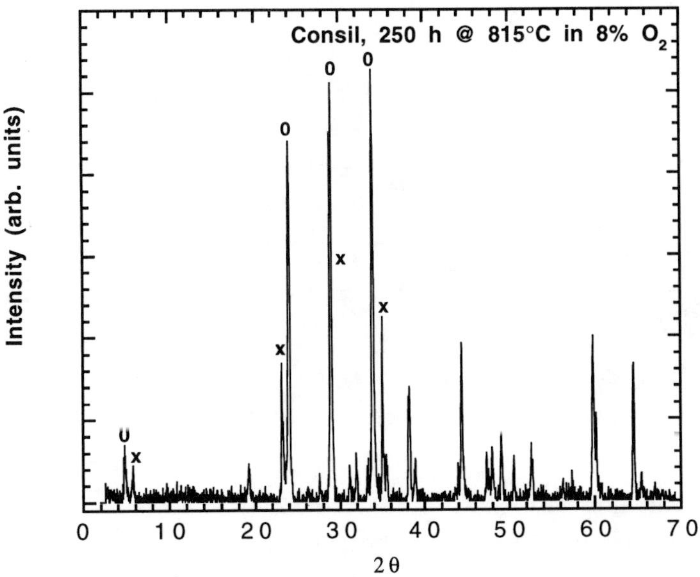

Fig. 2. XRD Patterns of Ag- and Consil-sheathed BSCCO tapes sintered for 150 and 250 h, respectively (0 = 2223 phase; x = 2212 phase).

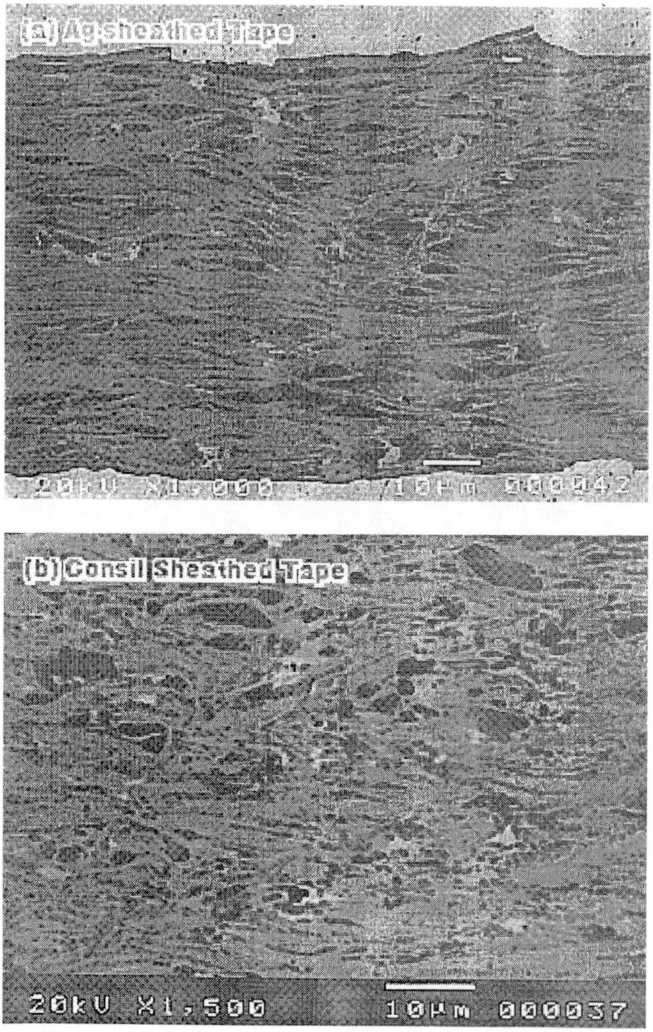

Fig. 3. Backscattered electron micrographs of (a) Ag-sheathed and (b) Consil-sheathed BSCCO tapes. Longitudinal cross section of Ag-sheathed tape shows well-aligned Bi-2223 grains with few secondary phases, whereas Consil-sheathed tape contains several large secondary phases, mostly alkaline-earth cuprates, as shown by dark regions.

- Reaction between core and Ni/Mg in Consil sheath. It is possible that Ni, with a structure similar to that of Cu, can substitute Cu in the core and degrade the transport properties. Additional TEM work is necessary to confirm Ni substitution for Cu in the core.
- Poor texturing at the consil/BSCCO interface due to undesirable reaction and presence of oxide precipitates. Because the transport current is believed to be carried at the Ag/BSCCO interface, the presence of these precipitates can disrupt texturing at the interface, leading to decreased I_c.
- Poor diffusion of O_2 through the sheath. Oxygen transport through the sheath is primarily through grain boundary diffusion, and the presence of stable oxide precipitates (MgO, NiO) may affect diffusivity of O_2 through the Consil.

Mechanical Testing of Tapes.

The tapes were soldered to steel plates and tested to failure in tension in an Instron machine at RT and at 77 K. The tensile stress/strain tolerance measurement set-up is shown in Fig. 4. Figure 5 shows stress-strain curves of Ag and Consil tapes under tension at RT and 77 K, respectively. The yield strength of the Consil tape was ≈220 MPa, more than five times higher than that of the Ag-sheathed tape which was just over 40 MPa. Ductility of the Consil tape remained constant at <10% at RT and 77 K, whereas the ductility of the Ag tape increased significantly from ≈ 13 to ≈37% between RT and 77 K. The increase in ductility with decrease in temperature is due to the increase in the strain-hardening exponent of Ag as the temperature decreases. However, consil tapes, which are already hardened, did not experience an increase in strain-hardening exponent.

Figure 6 shows retained I_c versus applied tensile strain for Ag and Consil tapes. These tapes were sintered for only 50 h because the portable current source could not supply over 10 A. The initial I_c of the Ag-sheathed tapes was 9 A and the Consil-sheathed tapes was <3 A. The plot in Fig. 6 shows that the I_c of the Ag tape degrades more severely than that of the Consil-sheathed tape, which shows a gradual decrease. Moreover, the stress required to produce an equivalent strain in the Consil-sheathed tape (Fig. 5) is more than four times higher than in Ag tapes, proving that the Consil-sheathed tapes are more stress tolerant than Ag-sheathed tapes. This makes them particularly useful in applications that generate hoop stresses such as coils and magnets.

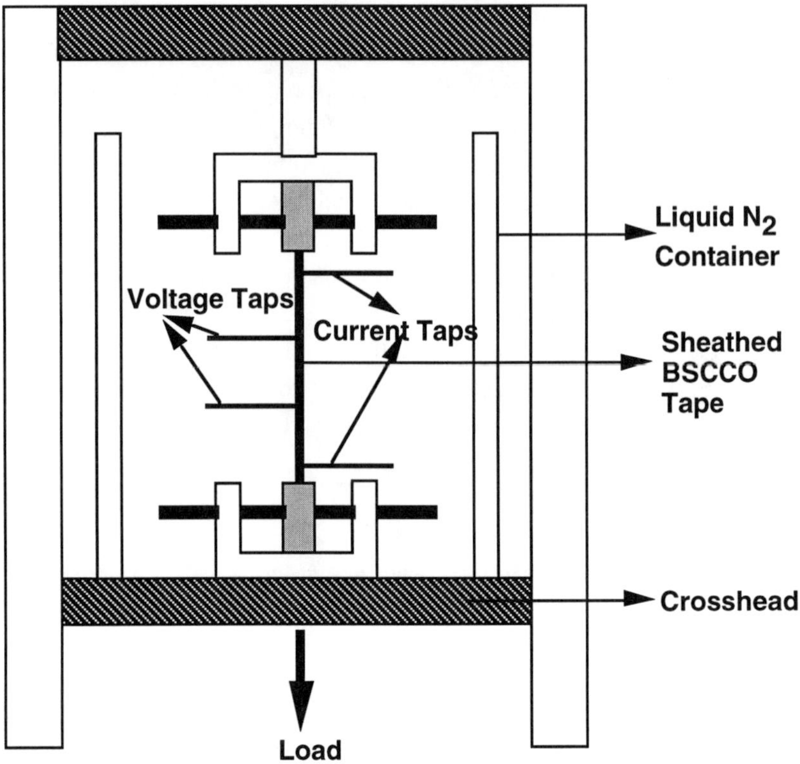

Fig. 4. Tensile stress/strain tolerance measurement setup. Strain is applied by an Instron machine. Test is interrupted at regular strain intervals to allow measurement of I_c values.

The apparently superior strain tolerance of the Consil tape may be due to interactions at the interface. However, these interactions also appear to be one of the factors that degrade the transport properties of the tape. To overcome the disadvantages of Consil, more compatible alloys must be used. We evaluated the strength behavior of precipitate-hardened Ag-1.2 at % Mg alloy (Ref. 4) and found that it has higher strength and is potentially more compatible with the superconducting core than Consil. Figure 7 shows a compressive stress-strain plot of one such alloy used in Ref. 4.

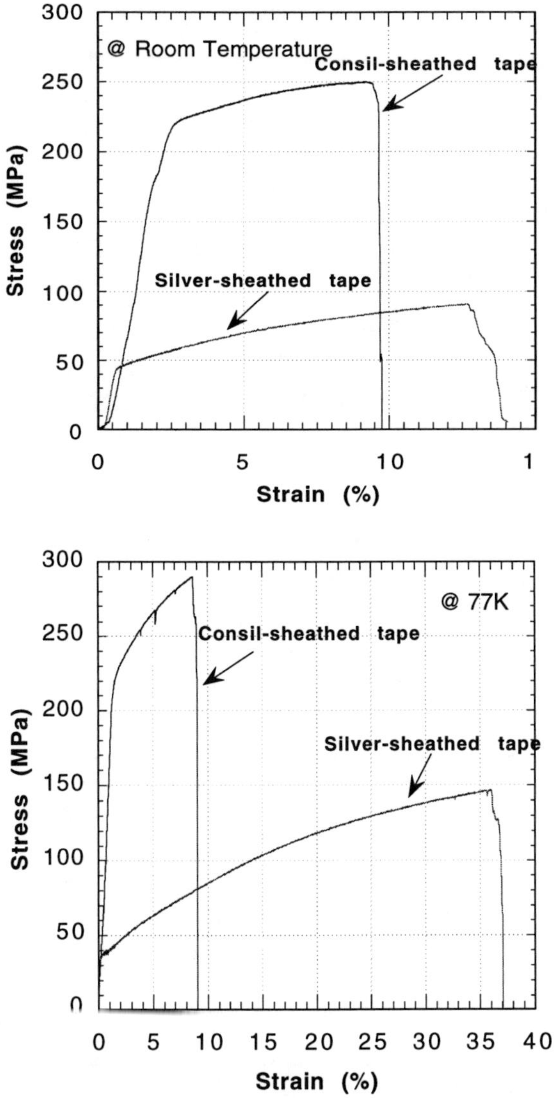

Fig. 5. Stress-strain plots of Ag- and consil-sheathed tapes at RT temperature and at 77 K. Ductility of the Ag-sheathed tape increases markedly from ≈13% at RT to ≈37% at 77 K, whereas the ductility of Consil-sheathed tape remains essentially unchanged.

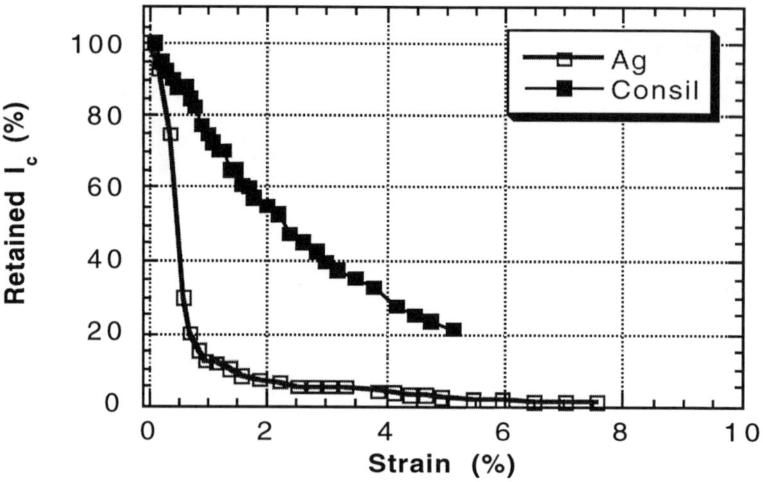

Fig. 6. Retained Ic (%) vs. applied tensile st.rain for Ag- and Consil sheathed tapes after 50 h of sintering

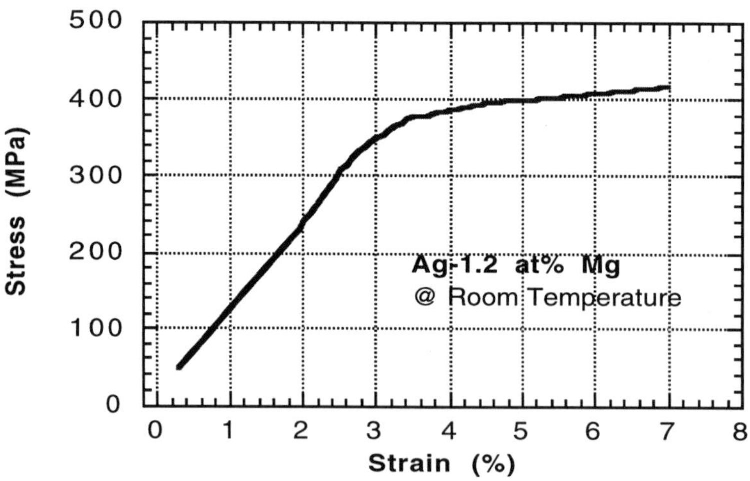

Fig. 6. Compressive stress-strain plot for precipitate-hardened Ag-1.2 at. % Mg alloy.

The compression strength of Ag-1.2 at.% Mg is ≈350 MPa, which is higher than that of Consil with a tensile yield strength of ≈120 MPa. Because the yield strength of metals is relatively the same under tension and compression, we have compared the compressive yield strength of Ag-1.2 at.% Mg to the tensile yield strength of Consil. The lower yield strength observed in Consil is due to the lower amount of alloying additions when compared to Ag-1.2 at.% Mg alloy. Since we believe that the Ni present in Consil could play a role in decreasing the critical current of the superconductor, the Ag-1.2 at.% Mg alloy devoid of Ni additions can overcome this disadvantage. Additional work is required to confirm the compatibility of this alloy as a potential sheathing material.

CONCLUSIONS

The strength and ductility of Ag- and Consil-sheathed tapes at room temperature and at 77 K were determined. The ductility of Ag increases with decrease in temperature, whereas the ductility of hardened Consil remains constant. The yield strength of the Consil-sheathed tape was more than five times higher than that of the Ag-sheathed tape.

Tensile strain tolerances of Ag- and Consil-sheathed tapes sintered for 50 h were evaluated; the results show an apparent improvement in strain-tolerance of Consil-sheathed tapes over that of Ag-sheathed tapes. However, the quality of the Consil tapes in terms of critical current is poor, and both microstructural and XRD characterization suggest that the superconducting cores in these tapes do not reach maximum phase purity even after extended hours of sintering. An alternate alloy of Ag, containing 1.2 at.% Mg (discussed in Ref. 4) could be a potential sheathing material if it can overcome the drawbacks observed in Consil.

ACKNOWLEDGMENTS

The authors thank S. E. Dorris, S. Mostovoy, and K. C. Goretta for helpful discussions. This work was supported by the U.S. Department of Energy, Energy Efficiency and Renewable Energy, as part of a DOE program to develop electric power technology, under Contract W-31-109-Eng-38.

REFERENCES

[1]Y. Feng, Y. E. High, D. C. Larbalstier, Y. S. Sung, and E. E. Hellstrom, Appl. Phys. Lett. **62** 13 (1993).

[2]U. Balachandran, A. N. Iyer, P. Haldar, and L. R. Motowildo, J. Met. **9** 54-57 (1993).
[3]M. Suenaga, Y. Fukomuto, P. Haldar, T. R. Thurston, and U. Wildgrüber, Appl. Phys. Lett. **67** 20 (1995).
[4]K. C. Goretta, M. Jiang, D. S. Kupperman, M. T. Lanagan, J. P. Singh, N. Vasanthamohan, D. G. Hinks, J. F. Mitchell, and J. W. Richardson Jr. IEEE Trans. Appl. Supercon., **7**, No. 2 (1997)
[5]J. Guo, J. A. Lewis, K. C. Goretta, and J. Schwartz, J. Appl. Phys. **78** 7 (1995).
[6]J. A. Parrell, S. E. Dorris, and D. C. Larbalstier, Adv. Cryo. Eng. **40** 193 (1994).
[7]J. Tenbrink, M. Wilhelm, K. Heine, and H. Krauth, IEEE Trans. Appl. Supercond. **3** 1 1123 (1993).
[8]Y. Tanaka, F. Matsumoto, and H. Maeda, IEEE Trans. Appl. Supercond. **3** 1 1161 (1994).
[9]J. P. Singh, J. Joo, N. Vasanthamohan, and R. B. Poeppel J. Mat. Res. **8** 10 (1993).
[10]J. P. Singh and N. Vasanthamohan, Accepted for publication at J. Mat. Res. (1997).
[11]H. K. Liu, Y. C. Guo, and S. X. Duo, Supercond. Sci. Technol. **5** 591 (1993).

EFFECT OF SHOCK LOADING ON CRITICAL CURRENT DENSITY OF BSCCO/Ag TAPES

S. J. Powell,[1,2] O.T. Inal,[1] and U. Balachandran[2]
[1]New Mexico Institute of Mining and Technology, Socorro
[2]Energy Technology Division, Argonne National Laboratory, Argonne, IL

ABSTRACT

Shock-loading has been used to introduce a variety of defects into superconducting BSCCO/Ag tapes. In theory, these defects will pin the movement of flux in these superconductors, allowing higher current densities to be passed through them. The tapes in the present study have been subjected to different peak shock pressures and pulse durations.

Critical current density (J_c) and magnetization measurements have been made on shock-loaded and pressed samples. The shocked samples have shown a degradation of transport current properties, but have maintained their high transition temperatures. It is thought that a postshock annealing to heal the microcracks in the material may improve the superconducting transport properties. It is also believed that the pressures to which these tapes were subjected completely saturated the material with defects. Investigation of the use of lower pressures may be advisable.

INTRODUCTION

Commercial application of high-temperature superconducting ceramics (HTSCs) depends on whether long lengths of the superconductor with high critical current densities (J_cs) can be fabricated. The obstacles to fabricating these superconductors have been materials problems, including grain boundary weak-links, flux motion, and poor mechanical properties not present in traditional low-temperature superconductors. These barriers are caused by small coherence lengths, high crystal anisotropy, low pinning energies needed to prevent magnetic flux flow and flux creep, and brittleness of the oxide ceramics.

The oxide powder-in-tube (OPIT) method has shown significant promise in the processing of long lengths of bismuth-based HTSCs, exhibiting J_cs of $>10^4$ A/cm^2 at 77 K. This method involves deformation processing of the oxides in a silver tube, increasing the texturing of the grains. The silver tube greatly enhances the mechanical properties of the oxide by providing a ductile confinement structure.

One problem that still remains for these bismuth-based HTSCs is that of flux motion at temperatures well below the transition temperature. When a current is applied, it induces a force, the Lorentz force (**J** X **B**), on these flux lines. If these flux lines are not pinned, they will move and dissipate energy, destroying superconductivity at temperatures as low as 30 K.

It has been shown that intragranular crystal defects pin flux lines and thus increase J_c. The optimal pin would have a dimension in the direction of the Lorentz force approximately equal to the coherence length, ξ, which is the dimension of the core of the flux line. In addition, to increase the volume of the flux line pinned, the defect should be much longer than ξ and lie parallel to the *c*-axis. Such defects have been introduced into superconducting materials through proton and heavy-ion irradiation, but such processing is expensive and the defects easily anneal out of the material.

It also appears that at higher temperatures, the flux lines begin to lose their integrity along their length, making flux motion easier. If the magnetic field is parallel to the *c*-axis, flux lines separate into "pancakes" within each Cu-O layer. These pancakes reduce the effective pinning volume and thus the energy needed for flux motion, thereby increasing the necessary number of pinning centers. A method for introducing a high-density, uniform network of stable defects is required. One such method may involve shock processing the material to introduce such a network. It is the purpose of the present work, therefore, to evaluate the use of shock processing to improve the utility of high-temperature superconducting ceramics.

EXPERIMENTAL PROCEDURE

BSCCO Tapes

Both monocore and multifilament bismuth-based tapes used in this study were made by the standard oxide powder-in-tube method. A final process annealing was performed on the tapes at 845°C for 100 h in air to promote conversion of the lower-T_c 2212-phase to the higher-T_c 2223-phase (1) and to heal any cracks remaining from the rolling process. Final cross-sectional areas of the monocore and multifilament tapes after the final annealing and before shock loading were approximately 8.76×10^{-3} cm^2 and 7.85×10^{-3} cm^2, respectively. To compare the shock-loaded tapes to conventional mechanically deformed tapes, both a monocore and a multifilament tape were cold pressed at 0.6 GPa between platens of tool steel.

Shock Assembly

The shock assembly used is illustrated in Figure 1. The containment, flyer, and momentum plates were all fabricated from a low carbon steel. The contacting surfaces of the plates were surface ground to provide an intimate contact between them. A thin layer of epoxy was used to adhere the momentum plate to the containment plate. The purpose of the momentum trap is to reduce the magnitude of rarefaction waves in the containment plate that would cause a tensile wave to propagate through the samples. The tapes were sandwiched between thin layers of lead in a groove machined in each containment plate.

The explosive used was Detasheet C1, a flexible high explosive made by Du Pont and consisting of RDX and an elastomeric powder. One set of tapes was shocked with one layer of C1, another set with two layers of C1, and a third set with three layers of C1. Variation in the number of layers allowed investigation of the dependence of defect saturation on variations in pulse duration and peak pressure.

The dynamic response of this assembly to the shock wave was modeled by a one-dimensional Lagrangian hydrodynamic computer code, MY1DL, based on difference equations by Neumann and Richtmyer (2). The calculational objective was to obtain an approximation of the pulse duration and pressure experienced by the superconducting tapes (Figure 2). Equation-of-state parameters for the explosive, steel, and lead were obtained from several different sources (3-5), but because such parameters were not available for the superconducting ceramic, the tapes were not included in the calculation. The shock impedance of the ceramic is significantly lower than that of lead; consequently, the actual pressure jump through the incident shock in the ceramic was lower than the calculated value in the lead. The estimated pulse duration is believed to be accurate. The pulse duration ranged from 0.9 μs for one layer to 2.2 μs for three layers of explosive. It is assumed that these short pulse durations lock a uniformly distributed array of defects into the material (6) and this, in turn, is expected to increase flux pinning and J_c in the superconducting filaments.

The temperature reached by the tapes is not known, but the layers of lead surrounding the tapes in the groove in the containment plate did not melt. Therefore, the temperature could not have been above the melting point of lead (327°C) for any appreciable amount of time.

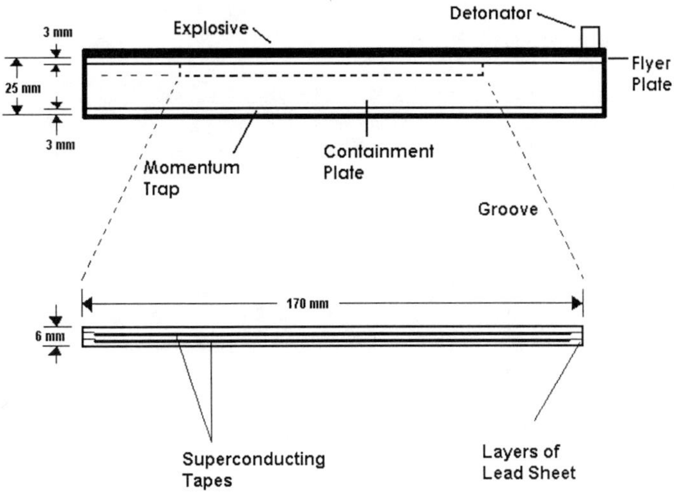

Figure 1: Explosive assembly for shock-loading superconducting tapes.

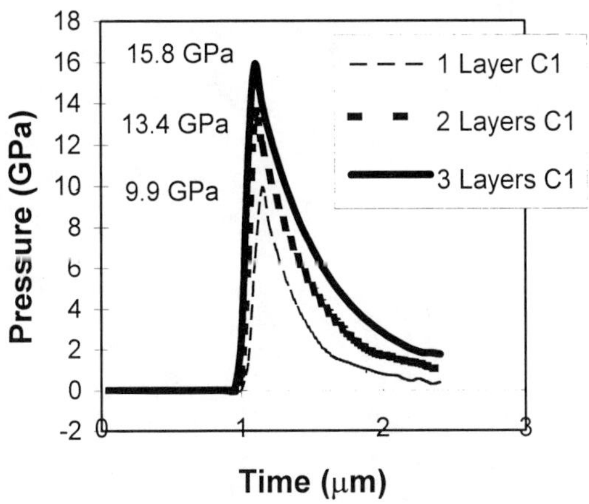

Figure 2: Pressure and pulse duration of various layers of C1.

Characterization

The superconducting tapes were examined by optical (OM) and scanning electron microscopy (SEM). Both cross-sectional and longitudinal samples were mounted in Lucite to investigate the microstructural features of both the monocore and multifilament tapes before and after shock-processing. Some of the samples viewed in the SEM were etched with a 1 vol.% perchloric acid - 99 vol.% 2-butoxyethanol etchant for 15 s to bring out the platy microstructure and were then sputtered with carbon to make the samples conductive. The cross-sectional area of each tape was measured by image analysis.

Transport J_c was measured by the four-probe DC technique. In this technique, the DC resistivity of the sample is measured by the voltage drop across a 1-cm span of a specimen with the passage of a known current. The critical currents of the samples were measured at 75 K with a 1 μV/cm criterion in a self-field. The J_c was then calculated by dividing by the superconducting cross-sectional area.

Transition temperatures of the samples were determined with a Quantum Design SQUID magnetometer. The tapes were cooled in zero-field to 10 K. They were then gradually warmed and the T_c was determined as the point at which the magnitude of the measured magnetization became zero. The volume of superconducting material was calculated from the measured magnetization. Magnetic hysteresis loops were formed by measuring magnetization in a varying applied field from 5 T to -5 T and back at both 10 K and 30 K with a 6-cm scan. Fields were oriented parallel to the c-axis.

EXPERIMENTAL RESULTS AND DISCUSSION

OM and SEM

The monocore samples were examined by OM and SEM (Figures 3 and 4).

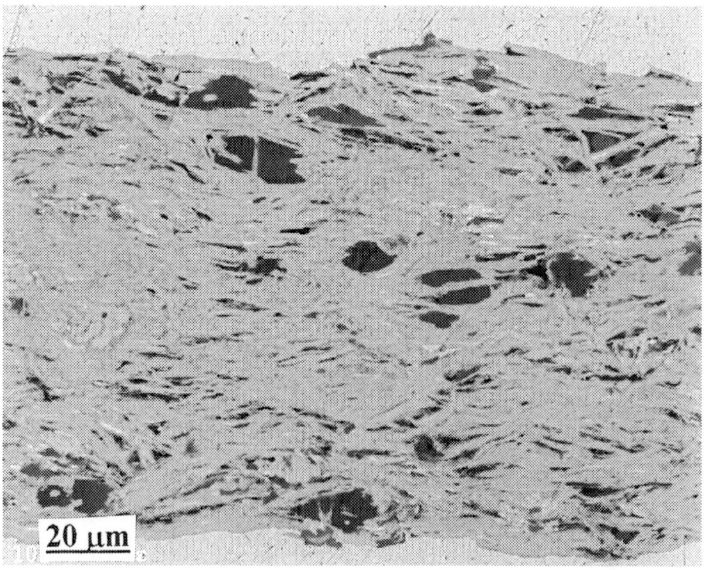

Figure 3: SEM photomicrograph of monocore sample before shock processing.

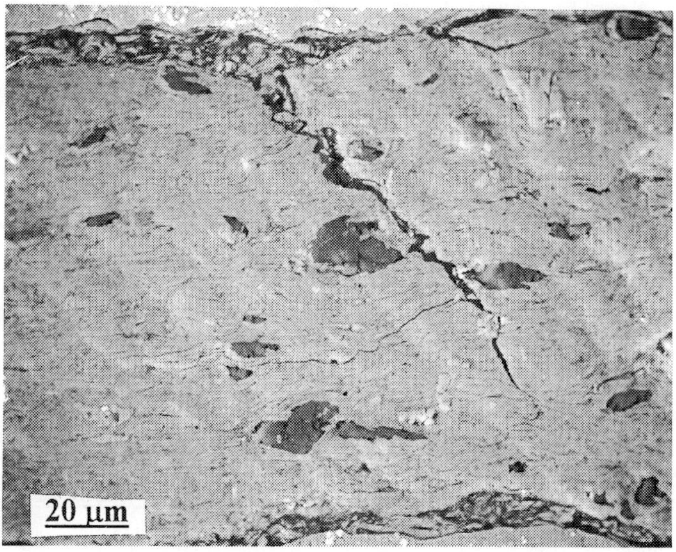

Figure 4: SEM photomicrograph of monocore sample shocked at 9.9 GPa.

Average thicknesses of the samples were measured, and both the monocore and multifilament tapes exper-ienced a reduction in thickness when rolled or shocked. The preshocked monocore thickness decreased by 10 µm (from 200 µm to 190 µm), while the average multifilament thickness dropped from 220 to 210 µm.

The percentage of ceramic in a cross section was determined by using an image analysis program from a montage of SEM photomicrographs. The average percentage of ceramic core in the monocore tapes was 44%. The multifilament tapes averaged 26% ceramic, the remainder of the tapes being silver in both cases.

The SEM photomicrographs showed no constriction, or sausaging, of the samples after shock processing (Figure 4). Preshocked monocore tapes contained porosity oriented parallel to the rolling direction between the platy grains and around second phases (Figure 3). Less of this type of porosity was seen within the ceramic in shocked samples, but there appears to be additional cracking at the BSCCO/Ag interfaces, as well as cracks running diagonally across some samples. Cracks at the interface may be a result of the large difference in impedance between the silver and the ceramic. The diagonal cracks tend to appear only in the vicinity of large (3-20 µm) nonsuperconducting second phases (the darker, more rounded inclusions) and are never seen to penetrate these particles. Such cracks could prove to be very detrimental to transport current densities.

Transport J_c Measurements

As mentioned previously, the transport current density of the monocore and multifilament samples were measured at 75 K in zero applied field with a 1 µV/cm criterion with the four-probe technique. Results of these measurements are listed in Table I as J_{ctrans}. This measurement is obtained by dividing I_c in the fourth column by the cross-sectional area of the superconducting ceramic.

Both of the preshocked monocore and multifilament samples were superconducting at 75 K. The shock process appears to have destroyed any transport current in the samples, most likely through introducing microcracks throughout the material. A J_c reading below 1000 A/cm^2 is generally considered not superconducting; most of the current is probably transported by the silver. This result is not surprising as other research has found decreases in transport J_c by at least a factor of 10 with samples pressed at only 2 GPa (7). Only the multifilament shocked at 9.9 GPa appeared to have survived the shock process, possibly because the additional silver within the superconducting ceramic improves its ductility and decreases the amount of microcracking by inhibiting the crack propagation.

Table I: Summary of electrical and magnetic measurements.

Sample	T_c (K)	T_{cMid} (K)	I_c (A) @ 75 K	J_{ctrans} (A/cm^2) @ 75 K	J_{cgrain} (A/cm^2) @ 30 K	Super-σ Volume (%)
Monocore Preshock	110	90.0	11.70	4255	4165	42.0
Monocore 9.9 GPa	102.5	68.7	0.200	130 *	590 *	6.7
Monocore 13.4 GPa	107.5	63.9	0.194	*	---	12.8
Monocore 15.8 GPa	100	65.5	0.188	*	---	10.3
Monocore Pressed 0.6 GPa	110	57.8	0.387	*	---	23.0
Multifilament Preshock	110	91.2	25.47	9580	9447	29.7
Multifilament 9.9 GPa	110	63.9	2.862	1080	1450	12.3
Multifilament 13.4 GPa	107.5	49.5	1.502	*	---	13.9
Multifilament 15.8 GPa	110	62.2	0.889	*	---	13.0
Multifilament Pressed 0.6 GPa	110	74.9	8.15	3060	---	22.5

*Does not appear to transport a superconducting current.

Magnetometer Measurements

Magnetization vs. Temperature: Magnetization of the preshocked, shocked, and pressed monocore and multifilament samples was measured as a function of temperature. The results can be seen in the graphs of Figure 5 and Table I.

Magnetometer measurements indicate a broadening of the transition temperature (increase in the difference between T_c and T_{cmid}), but a surprising discovery is that T_c was not altered by shock loading (Figure 5 and Table I). The magnitude of the magnetization decreased with shock loading, possibly indicating the introduction of a subgrain structure. Not much difference in magnetization is observed between the shocked samples, suggesting that defect saturation was achieved in all of the shocked samples. Analysis of the pressed tapes showed behavior similar to the shocked samples, but with less degradation of superconducting properties.

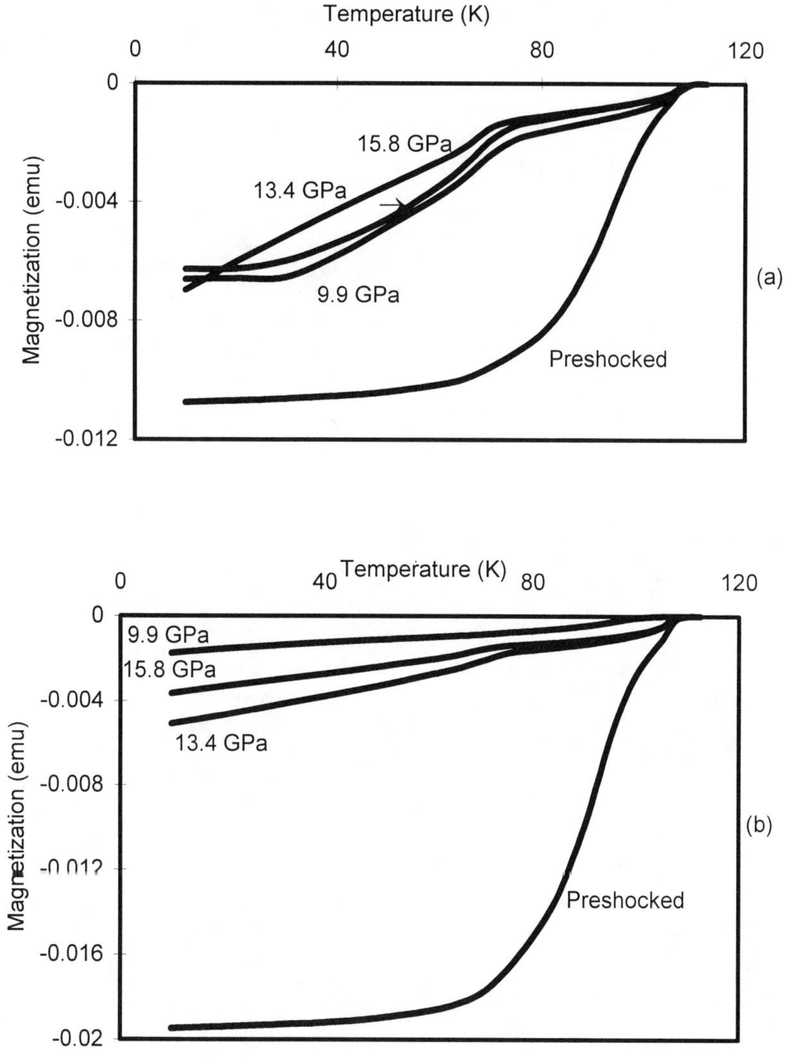

Figure 5: Magnetization vs. temperature for (a) monocore and (b) multifilament samples.

Total volume of superconducting material, calculated from the measured susceptibility with changing temperature, also decreased significantly upon shock loading (Table I). These decreases are most likely also caused by microcracks in the brittle ceramic material. It can be noted that the superconducting volume seems the smallest in the sample with the lowest shock pressure. Because the different shock pressures occurred in separate assemblies, it is possible that the assembly that was subjected to the lowest shock pressure experienced stronger rarefaction waves than the others due to a flaw in assembly, leading to higher tensile stresses in the samples and more damage.

Magnetization vs. Applied Field: Magnetization of the preshocked and shocked monocore samples was measured as a function of applied field. The results can be seen in the graphs of Figure 6

Calculation of J_c with the Bean critical state model modified for an orthorhombic sample was done for only the preshocked and 9.9 GPa shocked samples (Figure 6). Calculations showed higher values than the J_{ctrans}, but these values are not as significant because they do not report the amount of current traveling through the length of the material. However, they do help to illustrate that the cause of decrease in transport superconducting properties is most likely microcracks in the material and that the grains themselves have remained superconducting.

In a comparison of the hysteresis loops of preshocked and 9.9 GPa shocked monocore samples at 10 and 30 K, it is noted that at both temperatures the shocked samples exhibit a smaller hysteresis loop. In fact, the shocked sample at 30 K shows practically no loop at all. This decrease in loop size means that H_{irr} is very low and very little flux is being pinned within the sample.

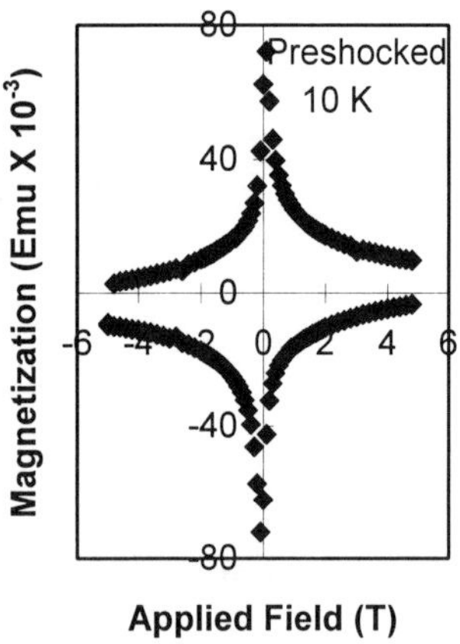

(a)

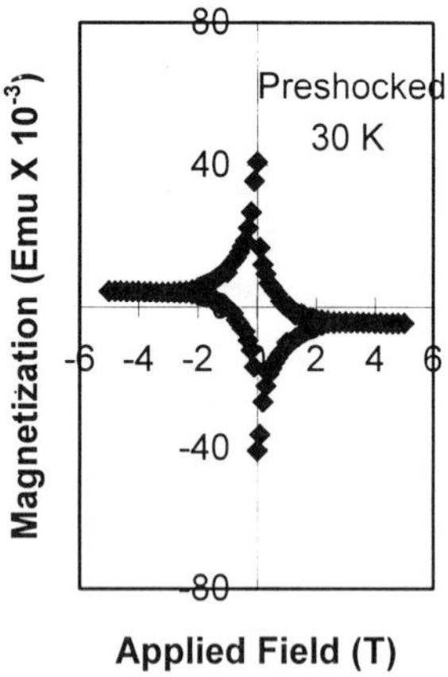

(b)

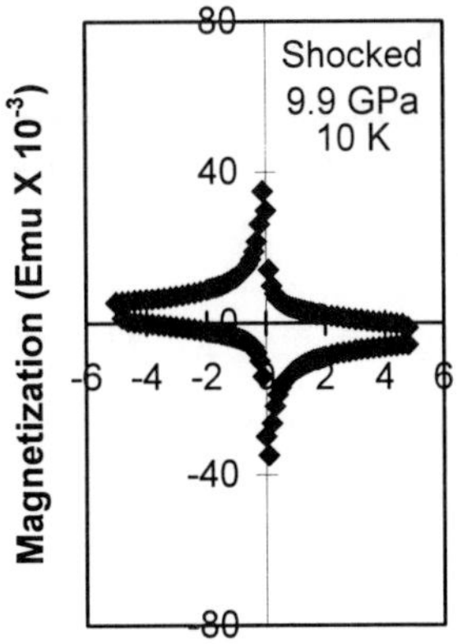

(c) Applied Field (T)

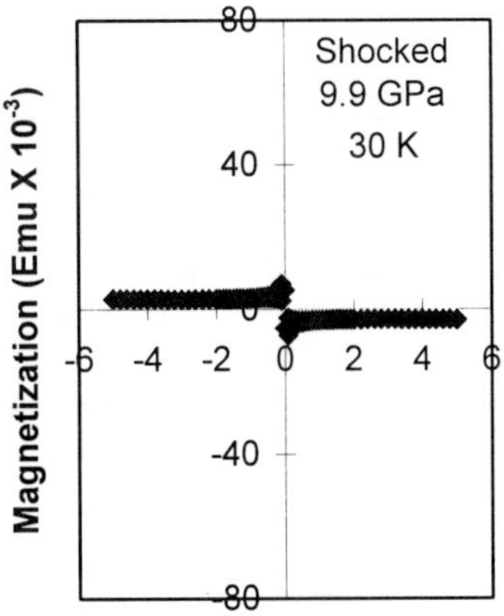

(d) Applied Field (T)

Figure 6: Magnetization vs. applied field for monocore samples: (a) preshocked at 10 K, (b) preshocked at 30 K, (c) 9.9 GPa shocked at 10 K, and (d) 9.9 GPa shocked at 30 K.

CONCLUSIONS

It is obvious that the samples subjected to shock processing exhibited significantly poorer superconducting properties, although the transition temperature changed very little. This lack of change in T_c implies that the oxygen stoichiometry of the material was not disturbed significantly by the shock process. This outcome illustrates that BSCCO is more stable stoichiometrically during shock processing than YBCO.

The degradation of transport current density after shock processing should not be surprising. Along with the possibility of introducing microcracks, considerable amounts of strain were introduced to the material. As more strain is added, the electron mean free path should decrease. This decrease would then lead to a decrease in the material's coherence length and an increase in the penetration depth. J_c is inversely proportional to the penetration depth of the superconductor; therefore, the more strain introduced, the lower the J_c. Because the cracks relieved stresses within the material, and therefore reduced the number of dislocations, pinning caused by these defects was not evident.

ACKNOWLEDGMENTS

We thank Jeffrey O. Willis and particularly J. Yates Coulter of the Superconducting Technology Center at Los Alamos National Laboratory for their help and assistance in measuring the electrical and magnetic properties of these samples. We are also grateful to Tony Zimmerly and the staff of the Energetic Materials Research and Testing Center for their assistance in performing the shock-loading. We would also like to acknowledge the help of J. L. Wise, D. Drumheller, and J. A. Voigt of Sandia National Laboratories. This work is supported by the U.S. Department of Energy (DOE), Energy Efficiency and Renewable Energy, as part of a DOE program to develop electric power technology, under Contract W-31-109-Eng-38.

REFERENCES

1. M. Takano, et al., *Jpn. J. Appl. Phys.* **27**, L1476 (1988).
2. J. V. Neumann, R. D. Richtmyer, *J. Appl. Phys.* **21**, 232 (1950).
3. D. J. Steinberg, "Equation of State and Strength Properties of Selected Materials" *UCRL-MA-106439* (Lawrence Livermore National Laboratory, 1996).
4. P. G. Shewmon, V. F. Zackay, eds., *Response of Metals to High Velocity Deformation* (Interscience, New York, 1960).
5. S. P. Marsh, *LASL Shock Hugoniot Data* (University of California Press, Berkeley, 1980).
6. M. A. Meyers, *Dynamic Behavior of Materials* (John Wiley & Sons, Inc., New York, 1994).
7. M. Daumling, J.-C. Grivel, B. Hensel, R. Flukiger, *Physica C* **219**, 429 (1994).

INTERFACE MODIFICATION BY BROMINE IN THE PROCESSING OF SUPER-CONDUCTING COMPOSITES

Sharmila M. Mukhopadhyay[1] and Chao Wei[2],
[1]Mechanical & Materials Eng., Wright State University, Dayton, OH 45435,
[2]Nautical Technology Corporation, 224 5th Avenue, New York, NY 10001.

ABSTRACT

An interfacial modification approach to the processing of Ag/superconductor composites has been investigated. It had been known earlier that even though Ag is the most suitable candidate for metallization of superconducting cuprates, it does cause detectable chemical degradation of the interfacial Cu^{2+} ions as observed by Photoelectron Spectroscopy. An investigation into possible processing steps to suppress this degradation has led to the observation that simply doping the $Y_1Ba_2Cu_3O_{7-x}$ (YBCO) surface with pure, dry bromine prior to contacting with Ag is perhaps the most effective way of accomplishing this goal. These results raise a number of scientific questions about the role of Br on the YBCO/metal interface. It is seen that Br changes the chemical states of the surface ions in YBCO and causes segregation of cations. These alterations of the surface composition may lead to the observed changes in interfacial chemistry. Most importantly, this simple step of halogen doping appears to increase the critical current density of the resulting Ag/YBCO composite by more than an order of magnitude. The feasibility of using halogen dopants as a simple, yet effective step in improving electrical transport at superconductor interfaces has been discussed. On a wider perspective, these results point to the significance of interface-property relationship of composite materials.

INTRODUCTION AND BACKGROUND

The interface between the metal and the superconductor is often the vital region that determines the behavior of the composite structure (whether it is a bulk conductor or a thin film device). However, this remains the least understood region because it is a discontinuity and, therefore, an imperfection. Several investigators have focused on the interfacial chemistry and contact resistivity aspects of metal-superconductor interfaces (1-6) and the general conclusion is that the very reactive metals (such as Ti & Cr) corrode the superconductor whereas non-reactive metals such as Ag and Au form acceptable contacts. Silver, in particular, is the favorite of most researchers (7-13) because it allows oxygen diffusion and enhances flux pinning in addition to providing a good electrical contact. Moreover, it has been shown (14-25) that the addition of silver to copper oxide superconductors leads to improved mechanical properties (strength, fracture toughness, thermal conductivity, and thermal shock resistance).

Despite these advantages, it is recognized that pure silver does not wet these oxides and tends to cluster on the surface. Additionally, if one zooms down at the first few interfacial layers of the composite, there is clear indication of chemical reduction of the Cu^{2+} ions caused by the metal (1,5,24-26). Since the oxidation state of Cu is known to be a key player in the superconducting behavior of these oxides, changes in this state is expected to influence electrical transport adversely. In other words, suppression of this effect might improve the transport properties of the composite. Moreover, any step, which can cause such a change in interfacial chemistry, is expected to cause modification of interfacial energies and microstructure. It therefore appears that a very useful route for the composite industry to take would be to systematically investigate how one can modify the interfacial chemistry in a favorable manner (viz. suppress the metal-induced reduction of copper) and how such a change will influence the microstructure and transport behavior of the composite.

Some of these questions have been answered in this article and some necessary directions for future investigations have been outlined. It has been seen that the single most effective way of preventing surface degradation of the surface Cu^{2+} ions is halogen doping of the superconducting oxide prior to contacting with silver. Br had been known in the past to restore superconductivity in vacuum annealed (and therefore, degraded) copper oxide superconductors, and its role in the bulk $Y_1Ba_2Cu_3O_{7-x}$ (YBCO) lattice had been investigated (27-32). Surprisingly, no investigation had addressed the surface/interface specifically. Only recently, some studies were conducted in this laboratory (26,33-35) and a semi-quantitative analysis of the Br-doped YBCO surface is evolving. These new studies show that not only does this halogen prevent the Cu ions at the Ag/YBCO interface from chemical reduction, it also changes the bulk properties of the Ag/YBCO composite. Most importantly this dopant has the potential to increase the critical current density of the superconductor by more than an order of magnitude. These observations immediately raise the question of what atomistic role this dopant could be playing on the surface of the multi-cation oxide. Few results addressing this fundamental question have been shown which indicate that halogen dopants are responsible for redistribution of cations on the surface.

EXPERIMENTAL DETAILS

For surface and interface studies, single crystal $Y_1Ba_2Cu_3O_{7-x}$ (YBCO, crystal size > 5mm) was obtained from University of Washington and compared with polycrystalline material (grain size about 10μm) made in house. The latter was obtained by sintering pressed powder (commercially available 99.9% YBCO, Alfa Chemicals) in flowing oxygen. A two-step heating cycle (2 hrs. at 950 C and subsequent 12 hr. anneal at 675 C) was seen to give the best results. The critical temperatures of both materials were checked to be about 92 K from resistance-temperature measurements using standard four-point probe. Samples were cut into flat discs, mechanically polished to 0.05μm final finish and annealed again at 675 C in flowing oxygen to restore surface superconductivity and release strain. These were the starting surfaces for surface spectroscopy studies. It is interesting to point out that, as far as these results are concerned, the only difference between the two materials is that the latter is more prone to contamination, as estimated from the carbon and oxygen photoelectron peaks (33).

All metal deposition and analyses were performed in a dual chamber, X-ray Photoelectron Spectroscopy (XPS) system described elsewhere (36). For this study, it is sufficient to note that all spectra shown here were acquired using MgK_{α} radiation with an expected energy resolution (width of standard $Ag3d_{5/2}$ peak) of 1 eV. Data acquisition and analysis was performed using DS800 software. Silver evaporation was carried out in a sample treatment chamber attached to the main analysis chamber with a background pressure of 10^{-9} torr. Evaporation rates were adjusted by trial and error to be very slow (about one monolayer per hour). Estimated coverages were calculated by measuring the total area of the photoelectron peak in comparison with that of the pure metal and using appropriate electronic escape depths available in the literature (37). This value assumes uniform coverage; therefore it is reliable only for sub-monolayer coverages (which is the focus of this study) and, subsequently, for layer by layer growth. After each deposition, samples were transferred to the adjacent chamber for XPS without breaking vacuum.

Bromine doping was performed by connecting the sample to a gas line, which was evacuated, then backfilled with vapors from a flask containing triple-distilled Br_2. The vapors were allowed to react for a given time (5 minutes in this case) and then pumped down. The line was then backfilled with inert gas and the sample immediately transferred to the XPS system for analysis. In order to minimize atmospheric contact, a plexi-glass box was attached to the sample load lock chamber that could be continuously flushed with inert gas during sample transfers and treatments. Sample etching and cleaning were performed right in this box. Sample transfers outside this box would be done in a container backfilled with inert gas. Care was taken that all analyses were performed immediately after the required surface treatment. This prevented any degradation of the surface from prolonged exposure to air or to ultra-high vacuum.

The Ag-YBCO composites having 32 wt.% silver were made from silver powder (maximum particle size of 3μm and 99.9% purity), and YBCO powder (particle size of 2-6μm and 99.9% purity). One set of YBCO powder was exposed to pure, dry bromine vapor for 5 minutes as described. The second set of YBCO powder was left un-doped to be used as control. The powders were mixed intimately, sintered, ground again and finally pressed and sintered using a two step heating cycle in flowing oxygen (35). Rectangular bars were cut from these composite pellets, ground and attached to leads with high purity Ag paint. Resistance *Vs* temperature and voltage *Vs* current measurements were made by the four-point probe technique. The criterion of 1μV/cm was used to take the Jc value. Since the focus of this project was to identify the influence of Br on the material, care was taken to prepare and characterize Br-doped and undoped specimens simultaneously, under identical conditions.

RESULTS AND DISCUSSIONS

The influence of Br doping on the surface of pure $Y_1Ba_2Cu_3O_{7-x}$ (YBCO) shows substantial change in composition and chemical states of the substrate ions. It is seen that among the three cations that make up YBCO, Br influences the chemical coordination of Y^{3+} ions most severely. Figure 1 compares the Y3d photoelectron peaks from pure and Br-doped YBCO surfaces. The binding energy (BE) of the peak has distinctly changed, indicating a change in co-ordination. Any apparent shift in BE due to specimen charging is always

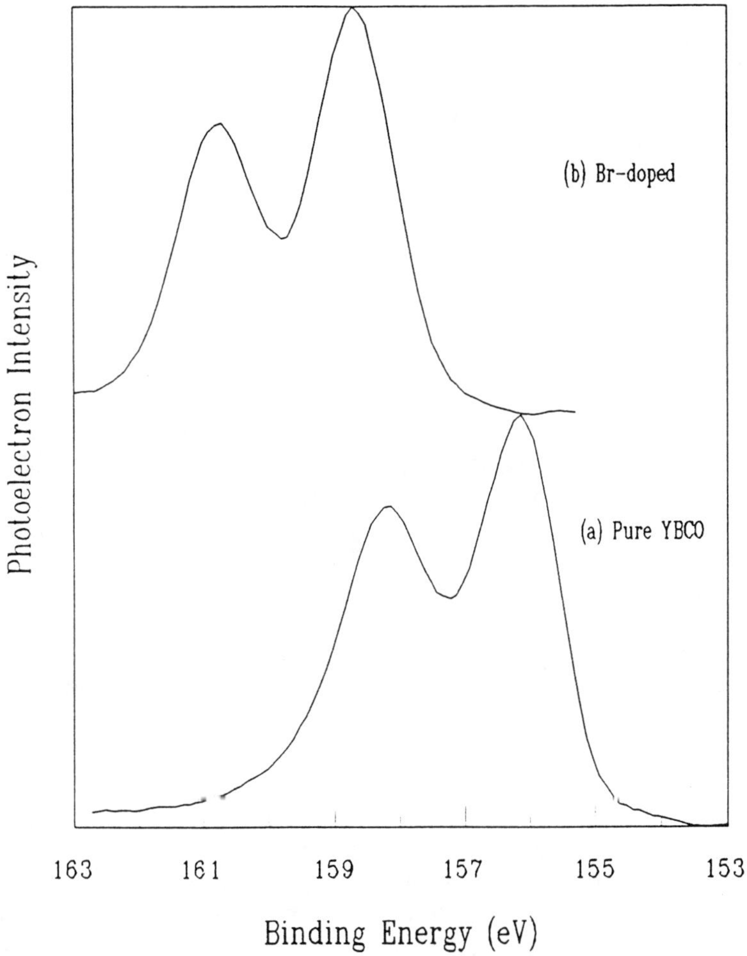

Figure 1: *Y3d photoelectron peak from (a) pure YBCO and (b) Br-doped YBCO crystal. The significant shift in binding energy indicates change in chemical co-ordination of the* Y^{3+} *ions (replacement of Y-O bonds with Y-Br bonds).*

compensated for by checking with adventitious carbon and other standard peaks as described elsewhere (36, 38); therefore this shift reflects a real change in BE. This change in co-ordination is accompanied by major enrichment of surface Y as seen in the following table.

TABLE 1
Photoelectron Peak Strengths of Cations
(Arbitrary units, normalized with total cationic peak strengths)

	Y	Ba	Cu
Undoped Surface	52	32	16
Br-Doped Surface	83	3	14

This table, which shows how the **relative** photoelectron peak strengths of the three cations change with doping, indicates that the surface is strongly enriched in Y and depleted of Ba when exposed to Br. The relative concentration of Cu remains almost unchanged. The O, Cu and Ba photoelectron peak shapes in this and related compounds are complex to begin with as discussed in the literature (32,33,38,39) and their shapes too are observed to change with Br doping. The most interesting observation is the severe depletion of Ba. Since there is no possibility of vaporization of any Ba-containing product from the surface, it appears that Br as a dopant causes Ba ions to move away from the surface. How much this reordering (movement of Y^{3+} towards the outer surface and that of Ba^{2+} away from it) has to do with the Br-affinity of the individual cations is a very interesting line of investigation that needs to be performed.

Figure 2(a) shows the Cu2p photoelectron peak obtained on pure YBCO. It has two components: the lower energy component is the main peak and the higher energy component is the satellite peak which originates from the paramagnetic state of the Cu^{2+} ion. It is therefore widely used as the signature for 2+ (or higher) oxidation state of Cu (40) and is known to decrease with chemical reduction of this ion. It was observed by us and by other groups (5, 26) that if Ag atoms are deposited on the YBCO surface, the satellite component is reduced indicating chemical reduction of substrate Cu^{2+} ions (Figure 2b). Figure 2c shows the Cu2p photoelectron peak from the Br-doped YBCO substrate that has the same coverage of Ag as does the undoped substrate

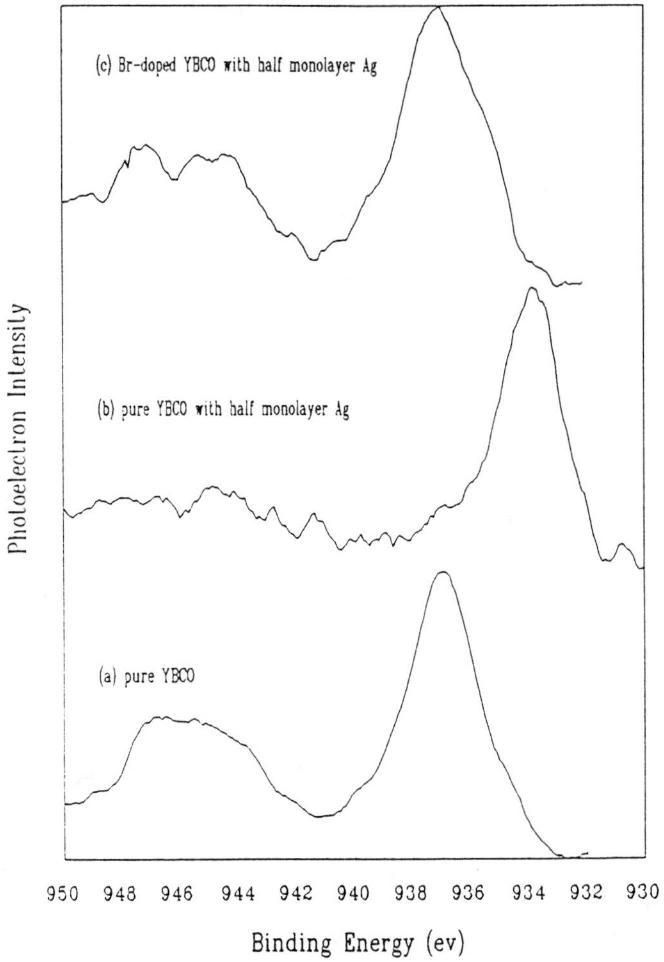

Figure 2: *Cu2p photoelectron peak from (a) Pure YBCO surface, (b) Same surface covered with half a monolayer of silver. Note the change in peak position and shape, especially the reduction of the satellite component. (c)Br-doped YBCO surface covered with half a monolayer of silver. The changes seen in Figure 2(b) are not apparent in this case.*

in Figure 2b. In this case, the satellite component is intact indicating higher oxidation state of Cu all the way to the interface.

Since the higher oxidation state of copper is definitely related to better superconductive behavior, any technique that can suppress this reduction may have desirable effects. This observation that doping the surface with Br_2 gas prior to metallization performs the task effectively leads us to believe that this line of investigation is not only of scientific value, but also of significant technological merit. Figure 3 shows a plot of the satellite peak as a function of Ag coverage in doped and undoped substrates. It clearly indicates that Br-doping is an effective method of controlling interfacial degradation of Cu^{2+} ions.

When a single dopant leads to changes in surface chemistry and interfacial bonding, the natural question which follows is: How would this change influence formation of the superconducting composite in terms of microstructure evolution and resulting electrical behavior? The microstructure of doped and undoped composites made from Ag and YBCO have been studied by both optical and scanning electron microscopes. These measurements, averaged over several samples and micrographs, have been tabulated in Table 2. It can be seen that the presence of Br leads to two changes: a significant increase in porosity and a moderate increase in particle size. This is an interesting observation that should definitely be linked to quantitative values of interfacial energies in the future. At this point, all that can be said is that the change in interfacial chemistry does lead to an observable change in microstructure evolution, which probably arises from changes in interfacial energies and mobilities.

The major practical impact of these investigations comes from the effect Br has on the electrical behavior of Ag/YBCO composites. Figure 4 shows the voltage versus current plot of the composites made with and without Br-doping. This data clearly indicates that Br has a significant influence on the critical current density of the composite: for the un-doped specimen, Jc is about $25 A/cm^2$ while that of the doped specimen is $350 A/cm^2$. It must be noted that these are totally untextured composites which are known to have very low Jc values in the anisotropic YBCO/Ag system. There have been a considerable amount of effort spent in the past on complicated texturing techniques (41-45) that improve current density, but none have focussed on the possible route of interfacial tailoring. It will be very interesting to see if incorporation of this step into other texturing techniques can boost their performance further.

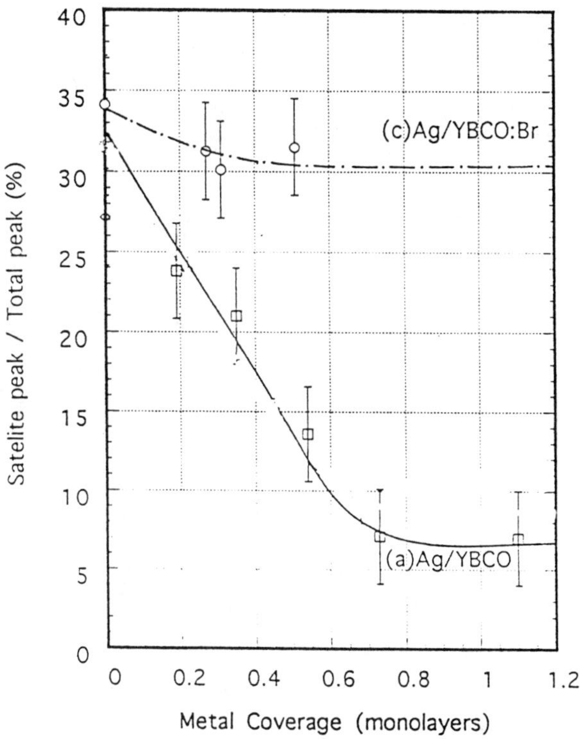

Figure 3. *Change in Cu2p satellite component (expressed as a fraction of total peak area) with metal coverage on the surface. (a) Ag deposition on pure YBCO surface and (b) Ag deposition on Br-doped YBCO surface.*

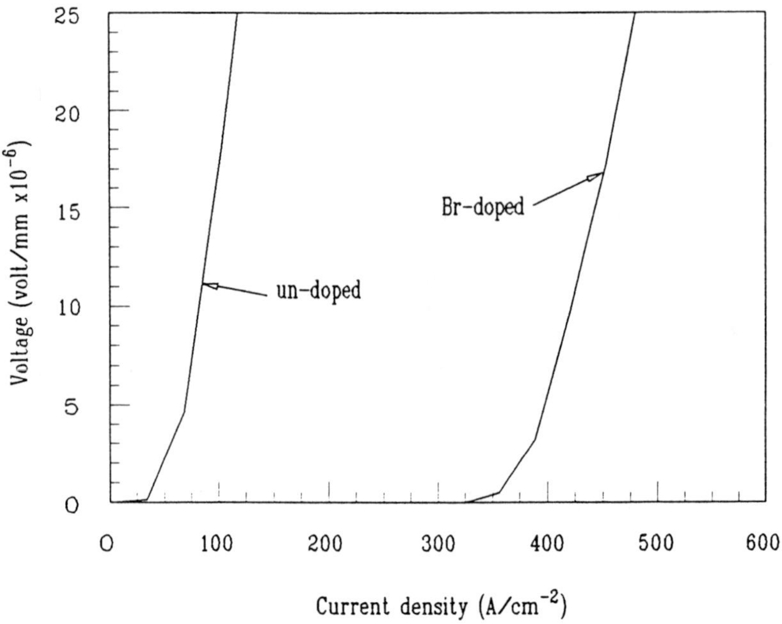

Figure 4: *Voltage Vs Current density (at 77K) of the two composites made with and without Br-doping. The boost in current density caused by Br-doping is very clear.*

TABLE 2
Comparison of Undoped and Br-Doped Ag-YBCO Composites

Sample	Average pore size (μm)	Porosity fraction (%)	Average size of Ag phase (μm)	AvgYBCO grain size (μm)	Jc at 77K (A/cm^2)
Un-doped	2.7	0.36	7.5	4.0	25
Br-doped	5.5	6.18	10.0	5.0	350

The observed improvement immediately shows that there is significant promise in pursuing this line of investigation. Br doping is a simple one-step-processing route that can be incorporated in any microelectronic fabrication facility. If it can change the interfacial chemistry to the level of causing significant improvement in Jc of the bulk composites, it might be able to boost electrical transport at all semiconductor-metal interfaces including those in VLSI devices.

CONCLUDING REMARKS

This paper points out the possibility of using the interfacial modification route for improvement in composite behavior. We see that the surface of the superconducting oxide $Y_1Ba_2Cu_3O_{7-x}$(YBCO) can be modified significantly by a beneficial dopant, Br. The halogen dopant leads to change in near surface composition and chemistry, notably, depletion of Ba^{2+} ions, enrichment of Y^{3+} ions and change in Y coordination indicating formation of Y-Br bonds. The change in the surface results in noticeable change of bonding property: When pure YBCO comes in contact with Ag, the interfacial Cu^{2+} ions are chemically reduced. But when Br-doped YBCO is contacted with silver, no reduction is observed. This change in interfacial interaction results in changes in microstucture development and electrical behavior of the Ag-YBCO composite. Most notably, the critical current density of the composite increases by more than an order of magnitude when the YBCO is doped with Br. This result is a clear indicator of possible advantages that can be gained in composite processing by interfacial modification investigations.

REFERENCES

1) J.H.Weaver, "Overlayer Formation in High-Temperature Superconductors", p210 in Interfaces in High-Tc Superconducting systems, Ed. by S.L.Shinde and D.A.Rudman, Springer Verlag (1993).
2) H.M.Meyer III, D.M.Hill, T.J.Wagener, Y.Gao, J.H.Weaver, D.W.Capone II and K.C.Goretta, Phys. Rev. B, 38, 10 (1988).
3) Q.Y.Ma, M.T.Schimdt, L.S.Weinman, E.S.Yang, S.M.Sampere ans Siu-Wai Chan, J.Vac.Sci.Technol., A9 (3), May-June (1991).
4) S.Jin, M.E.Davis, T.H.Tiefel, R.B.VanDover, R.C.Sherwood, H.M.O'Bryan, G.W.Kammlott and R.A.Fastnacht, Appl.Phys.Lett., 54, 25, 19 June (1989).
5) H.M.Meyer III, D.M.Hill, T.J.Wagener, J.H.Weaver, C.F.Gallo and K.C.Goretta, J.Appl.Phys. 65, 8, 15 April, (1989).
6) A.Plecenik, S.Benacka, M.Darula, S.Chromik, P.Mikusik & M.Grajcar, Solid State C0mm., 78, 9, 809 (1991).
7) A. G. Schrott, K. N. Tu, N. C. Yeh, G. Singco, A. Levi, and C. C. Tsuei, Phys.Rev., **B** 39, 2910(1989).
8) N. Savvides, A. Katsaros, and S. X. Dou, Physica **C**179, 3611(1991).
9) Q. Y. Ma, M. T. Shmidt, L. S. Weinman, and E. S. Yang, J. Vac. Sci. Technol., **A**9(3), 390(1991).
10) J. W. Ekin, A. J. Panson, and B. A. Blankenship, Appl. Phys. Lett., **52**, 331(1988).
11) T. W. Jing, Z. Z. Wang, and N. P. Ong, Appl. Phys. Lett., **55,** 1912(1989).
12) S. Jin, M. E. Davis, T. H. Tiefel, R. B. Van Dover, R. C. Sherwood, H. M. O'Bryan, G. W. Kammlott, and R. A. Fastnacht, Appl. Phys. Lett., **54,** 2605(1989).
13) B. D. Hunt, H. C. Foote, and R. P. Vasquez, Appl. Phys. Lett. **56,** 2678 (1990).
14) J. P. Singh, H. J. Leu, R. B. Poeppel, E. Van Voorhees, G. T. Goudey, K. Winsley, and D. Shi, J. Appl. Phys., **65**, 3154(1989).
15) J. P. Singh, J. Joo, N. Vasanthamohan, R. B. Poepel, J. Mater. Res., **8**(10), 2458(1993).
16) J. P. Singh, J. Joo, D. Singh, T. Warzynski, and R. B. Poepel, J. Mater. Res., **8**(6), 1226(1993).
17) J. Joo, J. P. Singh, T. Warzynsk, A. Grow, R. B. Poeppel, Appl. Superconductivity, **2**(6), 401(1994).
18) C. Vipulanandan, S. Salib, J. Mater. Sci., **30**, 763(1995).
19) T. Nishio, Y. Itoh, F. Ogasawara, M. Suganuma, Y. Yamada and U. Mizutani, J. Mater. Sci., **24**, 3228(1989).
20) D. Lee and K. Salama, Jpn. J. Appl. Phys., **29**, L2017(1990).
21) A. Matsumuro, K. Kasumi, U. Mizutani and M. Senoo, J. Mater.. Sci., **26**, 737(1991).
22) F. Yeh and K. W. White, J. Appl. Phys. **70**, 4989(1991).
23) G. Kozlowski, I. Marttense, R. Spyker, R. Leese, adn C. Oberly, Physica **C**173, 195(1991).
24) N. Savvides, A. Katsaros, and S. X. Dou, Physica **C179**, 261(1991).
25) M. T. Malachevsky, P. L. Villa, and L. Gherardi, Applied Superconductivity, **2**(1), 35(1994).

26) S. M. Mukhopadhyay, C. Wei, and T. C. S. Chen, Appl. Supercond. 3, 7, (1995).
27) Yu. A. Ossipyan, O. V. Zharikov, N. S. Sidorov, V. I. Kulakov, D. N. Magilyanskii, R. Nikolaev, V. Sh. Shekhtman, O. A. Volegova, and I. M. Romanenko, Pisíma Zh. Eksp. Teor. Fiz, 48 (1988)225 [JETP Lett. 48, 246(1988)]
28) Yu. A. Ossipyan and O. V. Zharikov, Physica C 162-164, 79(1989).
29) S. S. Rothman, J. Rothbaurt, and S. E. Baker, Phys. Rev. B40, 8852(1989).
30) H. B. Radousky, R. S. Glass, P. A. Hahn, M. J. Fluss, R. G. Meisenheimer, B. P. Bonner, C. I. Mertzbacher, E. M. Larson, K. D. McKeegan, J. C. O,Brein, J. L. Peng, R. N. Shelton and K. F. McCarty, Phys. Rev. B, 41, 16, 11141, (1990).
31) Y. Wang, A. M. Rao, J. G. Zhang, X. X. Bi, P. C. Eklund, M. S. Dresselhaus, P. P. Nguyen, J. S. Moodera, G. Dresselhaus, H. B. Radousky, R. S. Glas, M. J. Fluss, and J. Z. Liu, Phys. Rev. B45, 5 (1992).
32) Y. Fukuda, N. Sanada, Y. Suzuki, T. Goto, M. Nagoshi, Y. Syono, and M. Tachiki, Phys. Rev. B47, 418 (1993).
33) S.M.Mukhopadhyay and C.Wei, "Interaction of Br with $Y_1Ba_2Cu_3O_{7-x}$: its role as a dopant and an etchant", in review by Physica C.
34) C. Wei and S. M. Mukhopadhyay, Physica C 162-164, 356 (1996).
35) C.Wei, Ph.D. Thesis, Polytechnic University, (1997).
36) S.M.Mukhopadhyay and C.S.Chen, "Understanding Surface Activity of Oxides from Metal-Ceramic Interface Formation", J. of Vac. Sci. & Technol. A, 10 (6), 3545, (1992).
37) M. P. Seah, Surface and Interface Analysis, 9, 85 (1986) and Surface and Interface Analysis, Vol.1, 1, 2 (1979).
38) S. M. Mukhopadhyay and Tim C.S.Chen, J. Mater. Res., vol.10, No.6, June (1995).
39) C. C. Chang, M. S. Hegde, X. D. Wu, B. Dutta, A. Inam, T. Venkatesan, B. J. Wikens, and J. B. Wachtman, Jr., J. Appl. Phys., 67, 12, 15(1990).
40) F. Parmigiani and L. Sangaletti, Journal of Electron Spectroscopy & Related Phenomena, 66, 223-39, (1994).
41) M. J. Day, S. D. Sutton, and J. S. Abell, Cryogenics, **33**(1), 113(1993).
42) J. A. Xia, H. T. Ren, P. R. Munroe, Y. Zhao, H. K. Liu, C. Andrikidis, and S. X. Dou, J. Supercond., **7**(6), 947(1994).
43) C. S. Zhao, Y. B. Huang, G. K. Wang, Y. M. Wang, and F. X. Deng, Superconductor Sci. Tech., **3**(12), 587(1990).
44) T. Yamamoto, G. Terui, and T. Ueyama, Jpn. J. Appl. Phys., part 1, **32**(10), 4496(1993).
45) T. Kato, N. Yamada, and A. Imai, Physica **C**185, 2423(1991).

SUPERCONDUCTOR TRANSITIONS IN GLASSY SELENIUM

Semyon D. Savransky
West Coast Quartz Corporation
1000 Corporate Way, Fremont, CA 94539, USA.

ABSTRACT

The main superconducting parameters of glassy selenium have been analyzed in terms of the local pairs' theory. At pressures about 120 kbar the transition from the local pairs' regime to a BCS regime of superconductivity occurs in glassy selenium.

I. INTRODUCTION

In the past decade, it was theoretically shown that the maximum superconducting transition temperature occurs at the transition between the Bardeen-Cooper-Schrieffer (BCS) and the Bose local pairs (LP) mechanisms of superconductivity [1,2]. Apparently, an experimental proof of this idea has not been available. It seems reasonable that such a transition between Cooper's and Bose's pairs could occur in a systems where the local pairs [3] are created or destroyed by external fields. That is why glassy chalcogenides, in which existing local pairs could be destroyed by high pressure [4,5], attract our attention [6-9].

All elementary glassy chalcogenides are semiconductors or insulators in their normal - pressure form and they transform to superconducting phases under high pressure (P) and low temperature (T) [10-16]. The pressure dependence of the superconducting transition temperature T_c show the maximums for Se and Te to be below 10K [12,14].

In this paper we concentrate our attention on glassy selenium (g-Se) because its properties have been of considerable interest in recent years. However, some of its important parameters are not known at present. In order to clarify this problem we analyze the existing experimental data of g-Se for superconductivity property [10-16]. All reported experiments [10-16] show similar results. Among these data, Il'ina et al. Illustrated extensive details for g-Se [14].

In selenium the superconducting transition temperature T_c is higher in glassy samples than in crystalline ones, though the Debye temperature in the crystalline form is on the other hand higher than that in the glassy form [12,13]. This observation is very unusual for the conventional theory of Bardeen-Cooper-Schrieffer (see e.g., [17]).

At the pressures lower than P=110 kbar the values of T_c strongly increase with P, and at pressures higher than 125 kbar the T_c values quickly decrease as pressure increased. The maximum T_c occurs at P_o between 110 and 125 kbar. The pressure-induced crystallization occurs in g-Se at P above 130 kbar.

II. THEORETICAL BACKGROUND

It is known that the local pairs conception (e.g. [3,5, 18,19]) successfully describes certain properties of g-Se. For example, the negative effective Hubbard correlation energy, e.g. most of the peculiarities of the superconducting transition, is in qualitative agreement [6-9] with the theory of local pairs' superconductivity [1,2]. In the simplest version of this theory [18] the superconducting transition temperature T_c is connected with the Bose-Einstein condensation of the local pairs with the concentration N and mass M:

$$T_c = aN^{2/3}/M \qquad (1)$$

where $a = 6\cdot10^{-12}[K*cm^{+2}]m$, and m is the free electron mass. Coherence lengths Y may be calculated from the experimental data by the relation

$$dT_c / T_c \sim (R/Y)^4 \qquad (2)$$

where R is the characteristic local pair size and dT_c width of the superconducting transition.

Although the origin of attractive interactions between similar carriers in g-Se at high pressures has not been clarified as yet [4,5], we believe that the local pair's size sometimes increases under high pressure influence. We use R = 10 Å in this paper, which coincides with the typical distance of the medium-range order in g-Se as was observed in high-pressure structural experiments by Tanaka [4]. Note that equivalence of R to the medium-range order distance was also observed for other chalcogenide glasses [9].

The coherence length in the local pairs' theory corresponds to the thermal wavelength of local pairs and their concentrations are given by:

$$N \sim Y^{-3} \qquad (3)$$

The upper critical magnetic fields H_{c2}, the Landau-Ginzburg parameter K and the magnetic field penetration depth Q are defined as:

$$H_{c2} \sim f/Y^2 \qquad (4)$$

$$Q = (M/\mu e^2 n)^{1/2} \qquad (5)$$

$$K \sim Q/Y \qquad (6)$$

where f is the flux quantum and μ is the absolute magnetic permittivity.

The dimensionless constant of the electron-electron attraction G, the actual frequency f of lattice, the polaronic shift E_p, and the local pair's band with width t are defined from superconductivity parameters as

$$G \sim (\ln M/m)^{1/2} \qquad (7)$$

$$f = M/m(T_c^3 dT_c)^{1/4} \ln M/m \qquad (8)$$

$$E_p \sim G^2 f,\ t \sim f(m/M)^{1/2} \qquad (9)$$

III. GLASSY Se: RESULTS AND DISCUSSION

The main parameters of g-Se have been determined on the basis of the superconductivity data analyzed in terms of the local pairs theory. We assume that the Frohlich and deformation interactions prevail in g-Se and have used the coefficient of 1.5 in the calculation of the electronic parameters.

All estimations determined at various pressures are presented in Figures 1 to 3. One can see that for g-Se with small coherence length Y the superconductivity still persists up to very large values of the H_{c2}. From the Landau-Ginzburg parameter K one can conclude that g-Se is a strong type - II superconductor with small band width. In comparison with the conventional superconductors [17], it is seen that in metallic glassy Se the concentrations of charge carriers are very low and their mass is practically equal to twice the electron mass (the Cooper pair 2m) at P_o, and they move in the band t about T_C.

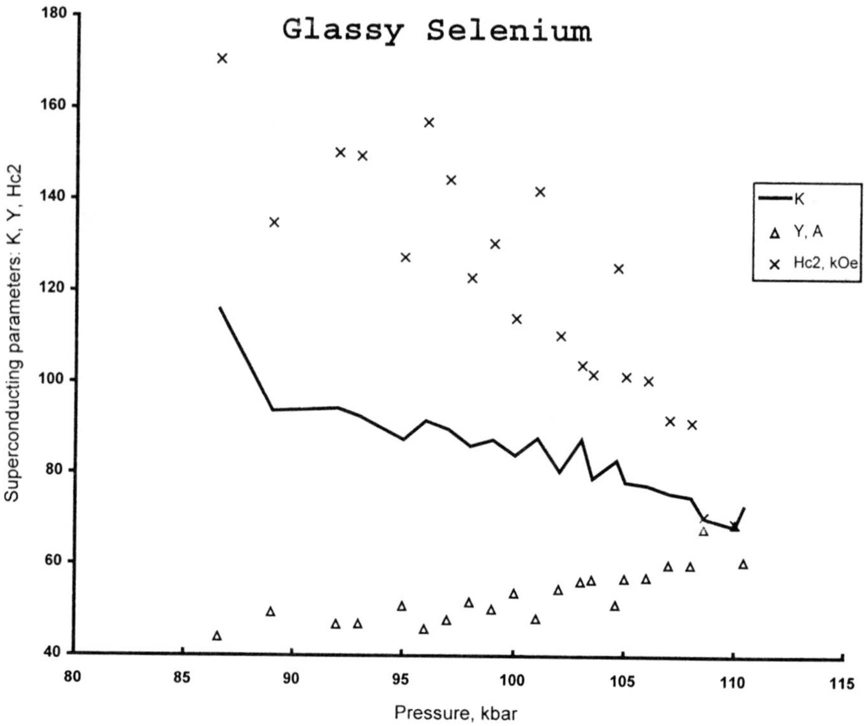

Figure 1. Pressure Dependecne K, Y, and Hc2.

The local pairs are self-trapped at soft atomic potentials having low elastic constants k in g-Se [19]. Growing pressure destroys these potentials with increasing

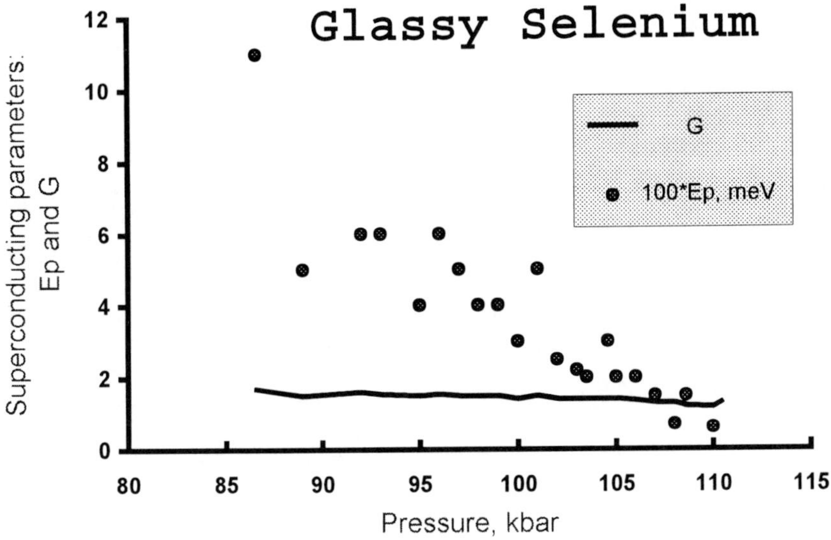

Figure 2. Pressure Dependence of Ep and G.

increasing k, and leads to decreasing E_p. Hence, the local pairs are destroyed by high pressures [5,9]. Furthermore, it was found that M suppression plays a major role in the T_c growth under P.

All estimations for the characteristics of g-Se obtained from experimental superconducting data [14] at moderately low pressures correspond to the predictions of the local pairs' theory of superconductivity. Some additional features of superconductivity in disordered forms of Selenium observed during the structural transformations are discussed in Ref. 20.

In the region where T_c decreases as pressure increases [14], effect of local pairs' suppression dominates and transition from the local pair theory to usual BCS mechanism [1,17] is possible. Therefore, glassy selenium as well as some other chalcogenide glasses could be used for investigation of this transition.

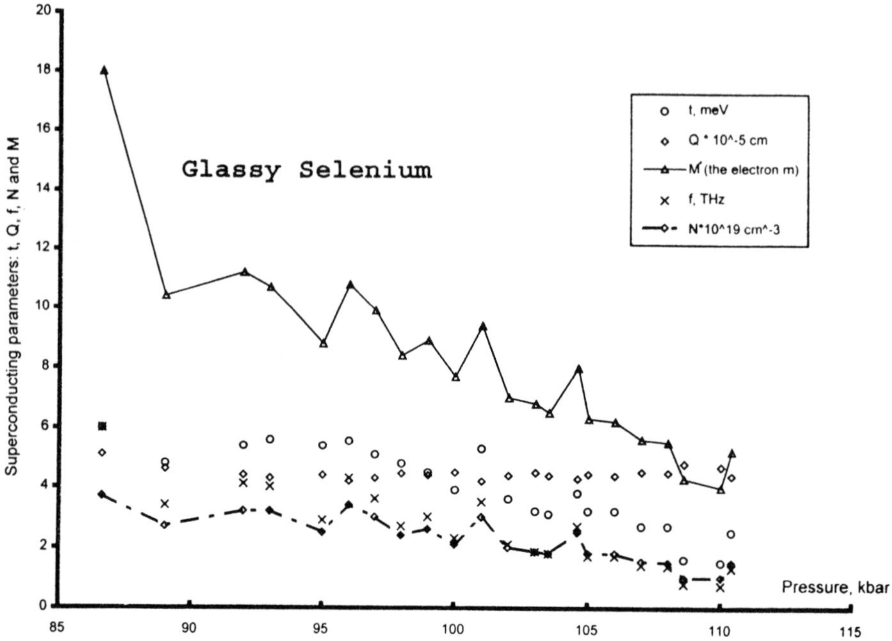

Figure 3. Pressure Dependences of t, Q, M, f and N.

ACKNOWLEDGEMENTS

The author wishes to thank Sir Nevill Mott and Prof. A. Alexandrov for useful discussions.

IV. REFERENCES

[1]R. Micnas, J. Ranninger and S. Robaszkiewicz, Rev. Modern Physics 62 (1990) 113.

[2]A.S. Alexandrov, N.F. Mott, Rept. Progr.Phys. (1994) 1197.

[3]N.F Mott and E.A. Davis, *Electronic Processes in Non-Crystalline Materials*, Clarendon Press, Oxford (1979).
[4]K. Tanaka, Physical Review B42 (1990) 11245.
[5]S.D. Savransky, In " The Hubbard Model : Its Physics and Mathematical Physics" *Eds. D. Bbaeriswyl, D.K. Campbell, J.M.P. Carmello, F. Guinea and E. Louis.* Plenum, New-York-London, 1995, p. 373-376.
[6]S.D. Savransky, J. Non - Crystal.Solids 83 (1986) 235
[7]S.D. Savransky, Solid State Communications, 79 (1991) 843.
[8]S.D. Savransky, J.Non - Crystal. Solids 164-166 (1993) 1173.
[9]S.D. Savransky, Phys. Rev. B52 (1995) 7434.
[10]J. Witting, Phys. Rev. Lett. 15 (1965) 159.
[11]F.P. Bundy and K.J. Dunn, High Pressure Science and Technology. Proc. 7th AIRAPT Conf., Oxford. 1 (1980) 461.
[12]A.R. Moodenbaugh, C.T. Wu and R. Viswanathan, Solid State Communication 13 (1973) 1413.
[13]F.P. Bundy and K.J. Dunn, Physical Review B22 (1980) 3157.
[14]M.A. Il'ina, Soviet Physics. Solid State 26 (1984) 5747.
[15]I.V. Berman, N.B. Brandt, Soviet Phys. J. Low Temperature Physics 16 (1990) 1227.
[16]Y.Akahama, M. Kobayashi and H. Kawamura, Solid State Communic. 84 (1992) 803.
[17]A.A. Abrikosov, Fundamentals of the Theory of Metals, North-Holland, Amsterdam (1988).
[18]S.D. Savransky, Materials Science and Engineering A145 (1991) 109.
[19]M.I. Klinger, Phys. Reports 165 (1988) 275.
[20]S.D. Savransky, High Temperatures - High Pressures. 27/28 (1995/1996) 377.

KEYWORD AND AUTHOR INDEX

Aganwal, M., 13
Arendt, P.N., 83
Ashcom, N., 185

Balachandran, U., 157, 207
Bhalla, A.S., 139
Bi-2212
  tapes, 175
  wires, 175
Bi-2223, 185
  stress/strain performance, 195
(Bi,Pb)-Sr-Ca-Cu-O, 55
$(Bi,Pb)_2Sr_2Ca_2Cu_3O_x$, 105
$Bi_2Sr_2CaCu_2O_8$, 93
Bingert, J.F., 175
Bowman, K.J., 105
Bromine, 217
BSCCO 2212, 41
BSCCO/Ag tapes, 207

Capponi, J.J., 3
Cellular telecommunications, 129
Chaillout, C., 3
Chenavas, J., 3
Coated-wire-in-tube processing, 185
Cook, L.P., 41, 55, 71
Corpus, J., 13
Coulter, J.Y., 83, 175
Critical current density, 207
Crystal growth, 41
Crystallization, 41

Detrie, T.J., 93
Dielectric substrates, 139
Dogan, F., 23
Dorris, S.E., 185
Dowden, P.C., 83
Duc, F., 3

Electric power applications, 157

Firing, 117
Foltyn, S.R., 83

Gaines, J.R., Jr., 13
Gautier, E., 3
Glassy selenium, 231
Greenwood, W., 71
Groves, J.R., 83
Growth model, 31

High-pressure synthesis, 3
High-temperature superconductor-based components, 129
High-temperature superconductor films, 139
Hodge, J.D., 129
Holesinger, T.G., 175
Hults, W.L., 83

Inal, O., 207
Incongruent melting, 105
Interface modification, 217
Isothermal melt processing, 173

Khanna, G., 195

Lanagan, M.T., 195
Le Floch, S., 3
Loureiro, S.M., 3

Manthiram, A., 117
Melt processing, 13, 41
Melting, 71
Metal-bearing precursors, 93
Moon, R.J., 105
Mukhopadhyay, S.M., 217
Muller, H., 129

Pachot, S., 3
Particle size, 23
Pb-2223((Bi:Pb)-Sr-Ca-Cu), 71
Pb distribution, 55
Peterson, E.J., 83
Pordel, P., 3
Powell, S., 207

Qu, D., 31

Ramanujan, C., 117

Reding, J.D., 23

Sandhage, K.H., 93
Savransky, S.D., 231
Seeded melt processing, 31
Sengupta, S., 13
Shi, D., 31
Shock loading, 207
Silver, 105
Singh, J.P., 195
Smith, J.L., 83
Sol-gel synthesis, 117
Solidification, 23
Superconductor, 157
  carbonate-based, 3
  composites, 217
  sheathed Bi-2223, 195
  transitions, 231

Tent, B., 31
Trumble, K.R., 105
Tsang, C., 117

Undercooling, 23

Vasanthmohan, N., 185, 195

Wei, C., 217
Willis, J.O., 83, 175
Wong-Ng, W., 41, 55, 71

XRD, 83

Y_2BaCuO_5, 23
$YBa_2Cu_3O_{6+x}$, 117
$YBa_2Cu_3O_{7-x}$, 23
$YBa_2Cu_3O_x$, 31
YBCO
  levitators, 13
  superconductor films, 83